T0176658

The Road to Quality Control

The Road to Quality Control

The Industrial Application of

Statistical Quality Control

by

Homer M. Sarasohn

Translated by N.I. Fisher & Y. Tanaka from the original
Japanese text published by Kagaku Shinko Sha
with a historical perspective by W.H. Woodall
and a historical context by N.I. Fisher

This edition first published 2019
© 2019 John Wiley & Sons Ltd

The right of N.I. Fisher, Y. Tanaka and W.H. Woodall to be identified as the authors of this translated work has been asserted in accordance with law.

Registered Offices
John Wiley & Sons, Inc., 111 River Street, Hoboken, NJ 07030, USA
John Wiley & Sons Ltd, The Atrium, Southern Gate, Chichester, West Sussex, PO19 8SQ, UK

Editorial Office
9600 Garsington Road, Oxford, OX4 2DQ, UK

For details of our global editorial offices, customer services, and more information about Wiley products visit us at www.wiley.com.

Wiley also publishes its books in a variety of electronic formats and by print-on-demand. Some content that appears in standard print versions of this book may not be available in other formats.

Library of Congress Cataloging-in-Publication Data applied for

ISBN: 9781119514930

Cover Design: Wiley
Cover Image: Homer M. Sarasohn Archives: IBM Corporate Headquarters, New Orchard Rd; Armonk, New York 10504; (914) 499-1900

Set in 10/12pt WarnockPro by SPi Global, Chennai, India

Printed in Singapore by C.O.S. Printers Pte Ltd

10 9 8 7 6 5 4 3 2 1

Contents – Summary

Translators' Preface

This book has unusual origins.

One of us (NIF) had the good fortune to form a close friendship with Homer Sarasohn in the last few years of his life. A few years after he passed away, in 2001, his daughter Lisa prepared his papers and other library materials for the Library of Congress. Some of his books were sent to his friends. Included in the resources sent to NIF was a small book written in Japanese but with the title also printed in English: *The Industrial Application of Statistical Quality Control*. The date of publication was given as 1951, making it a very early publication of any sort (let alone a book) on Statistical Quality Control, following Walter Shewhart's original publication. As such, it appeared to be a document of considerable historical interest, not least because of Homer's other work in Japan from 1946 to 1950.

Some time elapsed before permission was sought from Homer's family to arrange for a translation, which they kindly provided. Further delays have meant that it has taken nearly 15 years to bring the task of translation to completion. The translators also felt it important to provide some historical perspective for the book. This has been done by including two additional documents: one by an eminent academic researcher, Professor William Woodall, who kindly agreed to write an essay, positioning it in the published literature of the time, and the other a more general (previously published) article about Homer's work in Japan.

The background to the writing of this book is of some interest. In the early postwar years, Japanese engineers became aware of the widespread use of Statistical Process Control (SPC) in industry in the USA, and some perceived it as the driver for American manufacturing superiority.[1] However, Homer resisted the pressure to teach SPC until such time as he had had the opportunity to educate people in the more general principles of industrial management so that they would appreciate the context in which SPC should be deployed. He

---

1 Cf. Interview of Homer Sarasohn by Myron Tribus in 1988, http://honoringhomer.net/audio/interviews (accessed 19 July 2018).

and Charles Protzman conducted two courses (the so-called CCS courses) for managers in Japanese industry late in 1949 to provide this education. At this point, he was able to turn his attention to teaching Statistical Quality Control. Taking into account the comment about delays in publication in the Foreword to this book, we can infer that it was written in 1950. It had been his intention to run courses in quality control himself. However, in 1950 General Douglas MacArthur was reassigned to another task (in Korea) and took Homer with him. Consequently, Homer contacted Walter Shewhart, who was too busy to assist but referred him to W. Edwards Deming, who then visited Japan in 1950 for this purpose.

The translators wish to acknowledge some initial work done on the translation by Professor Shu Yamada and some of his students.

We have taken the liberty of correcting a few very small errors in the text. The diagrams have been re-drawn to reproduce the originals as closely as possible.

Finally, we note an irony that we feel would have amused Homer. As it happens, the last page we translated was the Foreword. We had assumed that Homer, who had taught himself Japanese, had written the book in Japanese. In fact, the Foreword reveals that this was not the case: it was translated into Japanese by an unknown team from Homer's original English script, so that we have, in effect, reverse engineered the book to approximate his original version.

N.I. Fisher
Y. Tanaka

The Road to Quality Control

The Industrial Application of

Statistical Quality Control

by

Homer M. Sarasohn

Original Japanese version published by
Kagaku Shinko Sha

Contents

Foreword

I owe Mr. Homer M. Sarasohn, the author of this book, very much for having taught me quality control. He had been working for the CCS of the GHQ. He is the person who proposed and executed an education program known as the "CCS course" for managers in Japanese industry. This book is his parting gift prepared on the basis of his belief that quality control is particularly important for Japanese industry. The main aim of his education program was to introduce quality control properly, in plain language, in a situation when there were few books available on quality control. Unfortunately, its publication was delayed for various reasons. Though it seems to me a little behind the times, I believe this kind of book is useful for readers because it is remains relevant. I regret there might be some parts that are not easy to read. It is because the translators have tried very hard to remain faithful to the original.

Mr. Homer M. Sarasohn now holds an important position in a company of management consultants named Booz, Allen & Hamilton, and is consulting for various large companies. Therefore, it may be said that this is a valuable book written by a first-class consultant in the USA who knows Japanese industry very well.

I believe this book will provide basic information to beginners about quality control and will stimulate people who have had experiences in practicing quality control by providing them with an opportunity for self-examination.

Finally, on behalf of the author, I thank those in the Japan Management Association for their efforts in making the translation.

Eizaburo Nishibori

I

Introduction

Preface

Product quality is an important matter in industry. It becomes more and more important given the recent trend that customers evaluate goods critically and make complaints about any issues that arise. Managers of companies, though they did not recognize it before, will now need to appreciate the following situation. In order for companies to be successful in modern society it is of fundamental importance for them to maintain the quality of products at the highest level possible, subject to operating profitably and producing products of uniform quality with reliable methods.

To realize this it is necessary to explain concretely to the managers of companies what management needs to do. The most important management principle for maintaining a high-quality product is to establish company-wide determination to serve their customers. After this is established and a firm decision is made to endeavor to maintain a reputation for high-quality products, management should inform all the employees of this strict requirement for product quality. Managers should take the initiative in related activities. They should show with their behavior that quality is their company motto, and they also should try to ensure that the workplace environment is conducive to maintaining the highest possible product quality.

The policies for the control and management of a company should be practical and informed by progressive ideas, and they should be adhered to. Analyze the market to obtain the information on customer needs, then design and produce the products based on the result of analysis so that production is profitable. Calculate and control the costs required for planning, production, and control, and avoid waste. Whether or not it is possible to maintain high quality really depends on the design and the administrative efficiency of the company. From the above standpoint, treatment and training of the employees is very important.

The Road to Quality Control: The Industrial Application of Statistical Quality Control by Homer M. Sarasohn, First Edition. Translated by N.I. Fisher & Y. Tanaka from the original Japanese text published by Kagaku Shinko Sha with a historical perspective by W.H. Woodall and a historical context by N.I. Fisher. © 2019 John Wiley & Sons Ltd. Published 2019 by John Wiley & Sons Ltd.

It will be difficult to maintain the required level of product quality without management and employees being in perfect agreement about the meaning of quality. Employees, including those working in factories and those working in offices, should be trained so that they are willing to cooperate with management. They should be trained to contribute to producing products of high quality and to understand that their efforts will result in profit not only to the company but also to themselves.

Since such a way of thinking does not occur naturally, management should take the initiative in training activities. If good plans are incorporated into the training programs for mental activities as well as for physical labor, the effects would spread into every area of the company. The basic thinking needed to improve the quality of output is common in the divisions related to physical labor and mental activity.

So far, we have discussed the problem of quality from the standpoint of management. Now we discuss it from the standpoint of operation. A reliable consistent production process and products with uniform quality can be achieved only when the production process is planned and controlled using scientific methods for the purpose of producing products of high quality.

Design, production method, and standards for materials and process specification should be appropriate, accurate and efficient, and they should be obeyed strictly. Workers should be trained and supervised so that they can understand and carry out their jobs well. Machines, equipment, instruments, tools, and skills of workers should be suited to carrying out the required tasks within the assigned time using appropriate methods.

Even if all the above conditions are satisfied, however, it is not generally easy to maintain product quality. This is because there is a shortcoming in the control system in general management. Recently, a lot of research has been done to address this shortcoming, and as a result some tools have been developed. They are statistical methods developed for the purpose of maintaining the quality of the production process. Using these methods, we can determine product quality with pre-assigned accuracy, and in addition we can improve the efficiency of production and also reduce costs. Scientifically speaking, the main aim of these methods is to control the factors that affect the quality of the final products going out to the market.

Establishment of a system of statistical quality control provides various advantages to the company as follows.

a. Variation of quality of the continuous production process is minimized.
b. Data based on samples taken from the production process provide reliable information to judge the state of the quality of the process.
c. Inspection cost (indirectly production cost) is minimized.
d. It is possible to obtain reliable basic information to test whether the current design and control limits for quality are appropriate or not.

e. It is possible to obtain practical basic information to assess the current production capability.

f. The results of sampling inspection usually provide proof of the actual quality of the products.

There can be no doubt that quality is the most essential foundation for a company to be successful. This has been demonstrated by many substantial companies. Quality is recognized as the most important matter. Therefore, it is necessary to position quality as the sound logical foundation for management and to adopt quality control as one of the valuable control methods among the industrial factors that are vital to the prosperity of the company.

There are two fundamental tasks necessary for successful practice of quality control. They are (1) to set up a system of control, and (2) to execute the actions resulting from the system of control.

The latter requires the support of management to ensure that the newly established system of control has a chance to work successfully; while the former provides the technical means, within which statistical quality control plays the principal role. We shall discuss these matters in detail below. There are some preliminary steps required to set up the initial control system. You should perform these logical steps one by one. However, as it is not essential to complete one step before proceeding to the next, it is sometimes convenient to perform some steps in parallel.

Step 1. Establishing the Quality Standard

Product quality captures the special characteristics useful for discriminating one's products from those of rival companies, special characteristics for discriminating among the same kind of products, or special characteristics for grading the products of the same production process. There are two purposes for quality. One is to demonstrate that two or more products are in the same category, as occurs when comparing the products manufactured by rival companies. The other is as a measure of fitness for purpose.

In either case quality is not measured on an absolute scale but merely on a relative scale. One can say that something is good from a certain point of view only when it is compared with some standard on a one-dimensional scale. In such cases quality becomes a variable and the result of a comparison can be expressed as quality being higher or lower (better or worse) than the standard.

When defining the quality of a product, possible characteristics related to the goodness of quality include size, materials, shape, chemical components, function, fitness for purpose, appearance, and practicability. These characteristics depend on (a) quality of design or (b) quality compared with the standard. The former relates to technical specification for manufacturing, and the latter relates to the method for comparing with the standard and the extent of difference from the standard.

Quality of design includes the problem of the trade-off between cost and outcome. It is a commercial issue to be considered carefully by both engineers and administrators. Consider an ideal case where engineers design a product with the best quality. If nobody buys the product in the market because of high cost, the idealistic engineers are to blame. If the engineers pursue high product quality simply for the sake of quality itself without any consideration for the practical use of the product, they are not fulfilling their duties. The engineers are responsible for the commercial aspect of the product as well.

If the accuracy of size, chemical purity, or any other quality attribute improves, the value of the product increases, but so does the cost. Eventually, comparing the additional value of the product that is due to an improvement of quality and the increase in cost required for the improvement, the point at which the latter exceeds the former will be the limit for quality improvement. Therefore, it is necessary for the engineers involved in production engineering and those involved in the production process to set the quality standard so that the standard is consistent with the reason for producing the product while the cost remains commercially viable.

The problem of quality versus cost discussed above also occurs in connection with quality compared with the technical standard. For example, suppose there is a production process in which 0.1% defective items are constantly produced. That is, on average there is one unit whose quality is below the required level among 1000 units produced in a day. It is necessary to inspect 1000 units per day in order to detect one defective unit. Ordinarily the inspection cost for 1000 units is much higher than the cost of the loss or inconvenience of a customer who buys the defective unit. In such cases it seems it is more economical not to carry out 100% inspection. It reduces the cost if we perform sampling inspection instead of 100% inspection, to gain assurance that the percent defective is less than 0.1%. Customers will be satisfied if an agreement is made in advance such that, when a customer happens to buy a defective unit, it will be replaced by a non-defective one or the possible loss will be compensated. The reason is that the cost required for 100% inspection causes an increase in the price of the product, which is not good for customers.

As the cost of quality is a vital factor in establishing the quality standard for the design or the technical standards, management should cooperate with the engineers to try to strike an appropriate balance between the cost of maintaining quality and the market price of the products. Among many factors that affect this issue, the main ones are as follows.

1. Desire of the manufacturer to maintain their reputation with their customers.
2. Quality of product design.

3. Skills of workers and quality of machines or tools needed to meet the prescribed quality requirements.
4. Research into target market, market needs for functionality, and requests from the market for practicability and price.
5. Rapid management capability to get access to raw materials.
6. Other economic factors related to achieving the required quality with reasonable production cost.

Step 2. Establishing a Section to Evaluate Quality

Having chosen the appropriate quality standard, it is necessary to establish a section to assess whether or not the produced products conform to the quality policy of the company. The assessment is made by inspectors in the company. Inspectors have duties to test the products against the quality standard set by engineers in the company. The relation between inspectors and the company is just like the relation between public prosecutors and a democratically elected government. In the case of government it consists of three independent divisions, *viz.* divisions of administration, legislature, and prosecution, for mutual inspection and maintaining balance. The administration practices the law which is made by the legislature, and the prosecutors check on whether the law is violated.

In a company, the administration division is represented by the chief officer and the general manager. For manufacturing, the duty of the legislature is practiced by the department of design. The duty of inspection is similar to that of prosecution: the inspectors judge whether the products satisfy the quality requirements by comparing them with the specification and standard decided by the department of design.

Of course, quality must be built into a product. It is not possible to change defective items to non-defective ones by inspection. The quality of products is not determined by inspection. An inspector's responsibility is simply to judge whether or not quality is of the required level.

Generally, responsibility is divided as described above, that is, the inspection department operates independently of the division of manufacturing. This is because in most companies it is desirable and necessary to do so.

Thus the three divisions, that is, the planning division for deciding on designs and standards, the manufacturing division for producing products based on the designs, and the inspection division for evaluating the quality level, are of the same rank and have their own independent functions.

If product quality is below the quality standard, it will be due to one or both of the following reasons: (1) design and specification made by the planning division are not good; or (2) the manufacturing division cannot produce products according to the design and specification, either because of a lack of capability or because of failures in the manufacturing process. It is important to keep the inspection division independent of the planning and manufacturing divisions in order to maintain the reputation of the quality of the products. When product quality is lower than the required level, the inspectors should point this out objectively, and clarify where the responsibility lies, with the cooperation of other appropriate sections. In fact, the inspectors should, to a certain extent, be independent of both of the design and manufacturing divisions in relation to stabilizing the company's product quality. In extreme cases, such as when defective items are being produced continually, the inspection division should have the authority to restrict the activity of the manufacturing division until the cause is found and removed. If there are any mistakes in the design of the product (or a component of the product) or in the system of manufacturing, the inspection division should be able to ask the design division to modify the design in order to remove the causes.

The main aims of inspection are not only to detect defects and remove them in the production process but also, more importantly, to prevent the production of defective items. Inspectors should have skills of a high level, and therefore the selection and training of inspectors is very important for the company. The reputation of the quality of the products – and therefore the reputation of the company – depend heavily on the skills, judgment, and honesty of the inspectors.

However, the rejection rate might become high owing to careless inspection or inaccuracy of measurement, even though both the design and production methods are not poor. It is not acceptable for inspectors who judge the quality of work done by other people to make mistakes in their own jobs. Inspectors should not make any mistakes because of a lack of attention to details, an irresponsible attitude in their daily jobs, or inappropriate or maladjusted machines or equipment.

Inspectors need to understand the importance of their role. They should be provided with the equipment necessary to carry out inspection, they should be aware of the necessity to interact objectively and cooperatively with people in other sections, and they should be trained well for their jobs. Of course, the manager also needs to have a very good understanding of the duties of the inspectors and to supervise them sensibly, so that if one of them happens to make a mistake in performing his job, the manager does not take him to task inappropriately.

Step 3. Establishing Inspection Standards

For production control it is necessary to start by establishing a series of inspection standards. The standards include identifying which objects are to be inspected; methods, instruments, and equipment to be used; conditions of inspection; the standard procedure for each inspection; the classification scheme for characteristics to be inspected; and the relative importance of inspected items.

Inspection standards consist of standards for (a) raw materials, (b) processing, (c) final products, and (d) their functions. The quality of purchased raw materials or parts or outsourced processed goods is checked for conformance to the inspection standards for raw materials. Inspection standards for processing are used to check characteristics such as shape, size, look, and finish of the intermediate products. Inspection standards for final products are used to check whether the final product matches the quality of design. When many parts are assembled into a large system the standards are used to check the functioning of the entire system.

These standards are written by the leaders of the design and engineering division or the production division, in accordance with company policy. The appropriate people to write standards describing the technical requirements will be chosen from those responsible for inspection. Inspection standards written in this way will be distributed to all inspectors, and they will be used just like the Bible.

Step 4. Selection of Inspection Methods

There are various kinds of inspection applied in industry. They include inspection of tools, inspection of the first product, patrol inspection, inspection of the process, concentrated inspection, and final inspection. Which of these varieties of inspection are applied depends on the purpose. Inspection of tools is used in the cases of frequently repeated operations to produce the same parts with the same equipment: for example an automatic stamping press machine. In such cases, the tools and forms are inspected for deterioration and damage rather than inspecting the components being produced.

Inspection of the initial product is also used in the cases of repetitious operations. The operation is checked completely for conformance to the requirements. If the first product conforms, it is assumed that subsequent operations should be all right. However, this approach will not be adequate in cases where most operations are manual and the operators are neither trained well nor supervised closely.

Patrol inspection is a method described by its name. Patrolling inspectors take small samples from the production process and check the quality of intermediate products.

Process inspection is what is most commonly applied. It takes place when proceeding from one process to the next process.

In contrast to the above inspections, which are made at the place of production, concentrated inspection is done elsewhere. For this method, since all inspectors work at the same place and therefore their work can be checked easily, less-experienced inspectors may participate and inspectors can share their work.

The final inspection is executed at the end of production, before sending products to customers or to the company warehouse, after all manufacturing processes and necessary treatments are over.

Inspection of tools and of the initial product can be regarded as types of sampling inspection, whereas for the other inspections either 100% inspection or sampling inspection is used. 100% inspection is used to remove defective items that might be produced during the production process, that is, to prevent defective items being sent out to customers. However, as is explained later, there are some factors that interfere with achieving this goal.

Sampling inspection is performed to obtain information about the production process, and its results are used for deciding on an action: accept, reject, or re-inspect the lot. It is a method to select and inspect part of a lot of products, analyze it, and judge the condition of the whole lot based on the result of the analysis. It is applicable in industry because it generally provides accurate results and is very economical to carry out.

Sampling inspection is based on a rule that a sample randomly taken from a group of products of the same kind represents the whole group within certain limits of probability. It is a reasonable assumption that, under the same conditions, the same outcome will occur. Given this assumption, a product produced by a certain method of production should be exactly the same as the succeeding product produced by exactly the same method of production. Similarly, the third and fourth products should be the same. Therefore, any product taken randomly from the group of products using the same method of production represents the whole group of products.

However, it is dangerous to assume the current production process in the factory is the same as that of the previous products produced. In general it is impossible to keep producing the products under exactly the same conditions. The attentiveness of workers changes as they tire. Machines and raw materials are not always of exactly the same quality. Therefore, even in the same manufacturing organization a series of products cannot be exactly the same.

The most that can be said is that they are the same, or homogeneous, within certain limits. The factors that affect the similarity of products such as the attentiveness of the workers and inhomogeneity of the raw materials are not the same or of the same level every day or every hour.

In other words, the quality of an item or a group of items produced at an arbitrary time point is a consequence of a combination of the conditions of all factors that affect quality. To assume that a single item represents all possible other items is to assume that no other combination of the conditions occurs in the other items. Since it is impossible to assume that a single item has all the information related to the random variation of all factors, it is necessary to make a selection of items to get a proper representation of a large lot. Each item should be selected using the same probability law with respect to the combination of the factors that affect the quality. Therefore, you have to study how to use random selection to be successful in the industrial use of sampling.

The next section provides a detailed explanation of how to use random selection. Here we simply add the comment that, whichever of these methods of inspection is selected, it has to be applied to appropriate points in the production process.

Step 5. Investigation of the Current State of Quality

To establish a system of quality control in a company requires a complete understanding of the current state of quality. For this purpose it is most effective to reinvestigate past production conditions, the design conditions in the engineering department, specifications for production, standards for inspection, and any other matters affecting the quality of work done by the production department. The reinvestigation should be done systematically to establish guidelines for production based on a scientific foundation, not just to obtain the information. A statistical investigation is one of the most useful methods for the above reinvestigation, in particular for use in detecting hidden causes of defects in production and variation in quality. The statistical investigation consists of collecting a series of facts related to errors of operation, delays in production in the factory, errors in design, standard costs, additional costs for control, and other factors affecting the efficiency of production. Such research studies have been carried out in many manufacturing companies, and as a result of analysis it is known that the main causes of defects in production can be classified as (1) bad raw material, (2) operator error, (3) low mechanical efficiency, (4) inappropriate design and (5) inappropriate methods of manufacture.

Inappropriate methods of manufacture generally derive from poor planning of operations, such as insufficient or no analysis of operations in the production line. They often occur because management does not standardize the training of workers, that is, it does not train them to work using the most efficient methods of operation as determined by the production engineers.

The problem of inappropriate design can be solved by ensuring that the design engineers have a deep understanding of economical methods of working and of economic issues. It is expected that the design engineers will themselves acquire the necessary theoretical knowledge and understand the actual capabilities and limitations of the workforce, and so be able to create a design that provides a sensible balance between scientific ideals and the reality achievable with the current equipment and skills of workers.

Low machine efficiency is a consequence of failing to maintain equipment in good condition. Machines should be cleaned, adjusted, and used carefully.

They should be checked regularly, and major problems should be prevented in advance by repairing small problems. Low machine efficiency is also a consequence of the failure of production engineers to analyze the production process carefully.

Operator errors occur because workers are not supervised properly, that is, they are not directed to investigate their operational errors. This happens when supervisors do not have a clear understanding of the need to supervise workers to ensure that they use proper methods, proper procedures, and proper tools, and when problems relating to the handling of raw materials, the order of the production process, and the assignment of workers with necessary skills to machines are not managed systematically.

Probably the most important factor resulting in defective products is the poor quality of raw materials. Poor raw materials and parts result in additional cost in the subsequent process and it is difficult to produce good final products. If the raw materials are bad, it is inevitable that the quality of the final products is low, and this results in higher production cost compared with the results from using good raw materials. Therefore, it is vital to make a careful check of raw materials in acceptance inspection based on the raw material specification. Materials should be inspected rigorously following the inspection standards at the acceptance stage and also during the production process. Bad materials must not be allowed to enter the production line.

There are at least two reasons, namely to produce products economically and to provide customers with high-quality products, why management has to ensure that arrangements are in place to obtain raw materials of high quality. An approach based on sampling inspection as described earlier can be used for this purpose. Its application is relatively simple and its effect is very substantial. The following shows some of its advantages.

1. It is possible to decide whether the whole lot of raw materials is to be accepted, rejected, or re-inspected by the inspection and analysis of a small sample taken from the lot. This method reduces the cost of inspection as well as the production cost.
2. It is possible for the company to make the supplier of raw materials aware of their quality by providing ongoing results of inspection of the raw materials. The company can obtain raw materials of high quality by requiring the supplier to maintain a high standard of quality.
3. It is desirable to record the result of analysis and the number of defective items in the sample for each lot of acceptance inspection. This allows their relationship to be investigated should any problems occur when using the lot of materials in the production process. It is useful in educating workers about the relationship between the quality of products and that of raw materials. The analysis of samples from acceptance inspection is useful in preventing quality problems and for predicting a possible increase in cost that

is due to quality problems. Since the supervisor of the production process knows the properties of the materials from the results of acceptance inspection, methods can be planned in advance to prevent possible quality problems.

4. The following storage policy is recommended. Store the lots of materials judged by sampling inspection to be of high quality in storehouses, and use those judged to be of relatively lower quality immediately in the production line. Then the production line works every day regardless of the result of the current inspection result, and materials of higher quality are accumulated gradually in the storehouses. As a result the requirement for quality improves gradually, and you do not need to be concerned about the possibility that the amount of production is affected because of the quality of materials, even if you send back the lots of materials rejected using the new standard to the supplier for exchange or adjustment.

5. Based on the results of analysis of sampling inspection, you can make a list of reliable suppliers of materials.

To remove the five causes of defect explained at the beginning of this section it is very important to detect the occurrence of a defect in the production process as soon as possible. A method developed in statistical quality control plays an important role in detecting the defect immediately. Quality control engineers are not themselves responsible for removing the detected causes. This is generally the role of production engineers. However, quality control engineers often have to make an initial assessment of the possible causes and point them out. Identification of the cause and developing an understanding of its relationship to special operating conditions can lead to significant improvement of the production process.

Statistical quality control aims to detect the causes of variation in quality logically and systematically using scientific methods. The methodology consists of relatively simple methods for detecting actual causes of defects, and by using it you can decrease the number of possible causes that may have a bad influence on production. How to use the methods in practice will be explained in succeeding chapters. However, before proceeding to the next chapter it will be appropriate to explain technical terms used in quality control and to introduce some fundamental concepts of theoretical statistics.

Technical Terms Used in Quality Control

1. Above normal
   A term meaning that the quality exceeds the standard.
2. Apparatus standard
   List of conventions including abbreviations and signs used in figures. It is used to describe the result of inspection.

3. Allowable defect number
   Allowable number of defective items per sample in sampling inspection.
4. Average quality
   Predetermined quality level with which observed quality is compared.
5. Band type chart
   Control chart to express the quality of products. It consists of three bands of "above normal", "normal", and "below normal".
6. Below normal
   A term meaning that the quality is lower than the standard.
7. Check inspection
   Sampling inspection of completed products carried out by the section responsible for final inspection.
8. Consumer's risk
   Probability that the lot with percent defective higher than the allowable percent is accepted.
9. Control chart
   Chart consisting of three lines, i.e. standard line and upper and lower control limits.
10. Control limits
    Upper and lower limits. The production process is said to be in control when the quality of the products plots within these limits.
11. Consumer's complaint
    Complaints from customers regardless of the type of consumer.
12. Defect
    Fault of a unit with a quality characteristic that does not satisfy the necessary condition.
13. Defective
    Unit with one or more quality defects.
14. Detailed inspection
    100% inspection of the units of the lot on characteristics or conditions related to quality.
15. End requirements
    Explicit or implicit requirements prescribed for the completed product.
16. Electrical inspection
    Inspection using an electrical test to check whether the item conforms to the requirement.
17. Engineering changes
    Changes that affect the function, appearance, maintenance, or interchangeability of the product.
18. Engineering change notice
    Notice of changes approved by the engineering department.

19. Engineering information
    Changes of technical specifications, charts, and their requirements approved by the engineering department.
20. Expected quality level
    Prescribed quality level or predicted quality level at the time of delivery to customers.
21. Final inspection
    Inspection of products to confirm that they satisfy the global requirements before delivering to customers.
22. Fraction defective
    Proportion of the units among the total units that are below the standard.
23. Inspection item
    Characteristics, performance, properties, and other things to be inspected.
24. Isolated defect
    Very rare defect, the possibility of improving which can be neglected.
25. Limit line
    Line indicating upper or lower limit of the control chart. The pair of allowable limits of variation.
26. Lot
    A certain amount of the same kind of raw materials or same kind of products taken from the same source. For example, in the case of producing products on a conveyer or on a production line a lot is defined by the amount of product produced per day.
27. Manufacturing information
    Documents concerning the method of manufacturing, such as design, specification, instructions, notice of changes, layouts, and other relevant information.
28. Manufacturing standards
    A set of approved directions, descriptions, and practicing methods to measure the gauges and sizes of the standardized components and to decide the tolerances and other procedures.
29. Mechanical inspection
    A form of inspection using mechanical tools or gauges.
30. Nonconformance
    A product or unit that does not conform to specification.
31. Normal distribution
    It is known that Gaussian probability distribution, also called the Normal distribution, generally approximates the observed frequency distributions of measurements. For the Normal distribution the interval between three times the standard deviation above the average and three times the standard deviation below the average contains 99.7% of the distribution.

32. Observational standard

    An object made as a visible standard for assessing quality characteristics, such as shape, color, composition, and appearance. Each item should conform to this standard. It is used in situations where it is difficult to describe the standard with pictures, numbers, or measurements and language.

33. Observed number of defects

    Number of defects detected in the sample.

34. Output

    Amount of product produced during a certain period.

35. Percent defective

    Percent of defective units among all units inspected.

36. Piece parts

    Parts constituting the components of the assembled product.

37. Process

    Operations conducted in the assembly of the products.

38. Process inspection

    Inspection of parts, partial assembly, and process.

39. Producer's risk

    Probability of having to re-inspect a lot conforming to specification because a sample is rejected as a result of sampling inspection. Here, "producer" means the company itself or a division of the company that supplies materials or parts to other divisions.

40. Quality level

    Average quality.

41. Quality report

    Periodic report to the management on the current state of quality.

42. Quality surveys

    Surveys to assess whether the production process conforms to the design and to evaluate globally everything that affects product quality, including the production method, the procedure and the results of inspection, and the level of experience of the workers.

43. Raw material inspection

    Inspection of raw materials and outsourced components.

44. Sample (lot)

    A certain number of units selected from the lot for sampling inspection.

45. Sampling inspection

    Inspection of the sample selected from the lot to decide whether to accept, reject, or re-inspect the lot.

46. Spotty condition (lot)

    Insufficient condition of quality concentrated in a small part of the whole group of products or of the lot.

47. State of control
    Production condition shown by the control chart when a sequence of 25 or more points plots within the control limits.
48. Subsequent lot
    Term used in a situation where successive lots produced in a certain production process are rejected by inspection.
49. Temporary information memorandum
    Memorandum issued for a certain period or a certain amount of production or to advise that it is permissible, on a temporary basis, not to conform to the standard instruction for production.
50. Lot tolerance percent defective
    Critical value of percent defective for the lot to be accepted, i.e. a lot with percent defective larger than this critical value should be rejected.
51. Tool made sample
    Sample of a new product which is made in the production process to be used in actual production before producing the products formally. It is used as an example of the new product.
52. Visual inspection
    Inspection to investigate appearance.
53. Workmanship
    Quality of product determined by tool, accuracy, and skill of worker.

II

Probability and Statistical Inference

In order to understand Statistical Quality Control properly, you need to know about statistical inference and Probability. In the theory of Probability, the probability of an event occurring is based on the hypothesis that states "the results occurring under the same conditions are considered the same". This is a fundamental law of Science: the belief that you'll always get the same results if you repeat the experiment under precisely the same conditions. However, in the case of products made in factories, it cannot be the same as what you do in the laboratory, where it is easy to set the same conditions. Although we try to control the operating conditions in several ways, such as managing the manufacturing process, doing process analysis, coaching people in standard operating procedures, or by trying other control methods, the products produced frequently don't meet the required standard because of a variety of problems caused by the bad effects of those trials themselves. Even if the manufacturing conditions at one time are consistent with those at another time, they are only "relatively the same". Therefore, it is difficult to avoid variation in the quality as it's impossible to equalize the characteristics of each one and, moreover, it is hard to do this in an economical way. However, we not only can, but we must, minimize, although not exactly, the amount of variability in the quality of products in a unit or lot, as we explain precisely in Chapter III. This is the purpose of quality control.

Let's explain this idea with an example. Suppose you strike a match. It ignites and produces some smoke. And you expect to get the same result if you strike another one; that is, it will also ignite and produce smoke. Now, you take out a match and when the second match actually flames up, you'll think that the conditions for the two separate actions were the same because they produced the same results, and you'll also think that all matches in the box will be of the same quality.

However, if the second match hasn't ignited, you would suspect that the operating condition was not the same as for the first match. Then again, what type of conclusion can you reach if the second one hasn't ignited but the third one has?

The Road to Quality Control: The Industrial Application of Statistical Quality Control by Homer M. Sarasohn,
First Edition. Translated by N.I. Fisher & Y. Tanaka from the original Japanese text published
by Kagaku Shinko Sha with a historical perspective by W.H. Woodall and a historical context by N.I. Fisher.
© 2019 John Wiley & Sons Ltd. Published 2019 by John Wiley & Sons Ltd.

(The area of friction is supposed equal and external conditions are unchanged.) All matches look the same and they are produced by the same method, from the same materials, and almost at the same time. They are produced to have the same characteristics for use but one of three was defective. Does this result show 33% of all matches are defective? One way to find out the truth is to strike all the matches and see what happens. Another way is to take a representative sample of matches, inspect them, and draw a general conclusion for all the matches based on the result.

It is obvious that in the first method no matches are left after doing the complete inspection. On the other hand, for the sampling inspection, there will be two questions of a probabilistic nature. The first question is how certain it is that all matches have the same quality. From simple experimentation we know that there are some differences between matches. Therefore, what we want to know is not how different they are but the probability they have the same quality to be usable.

The second question relates to the probability that the characteristics of the matches in the sample are consistent with the characteristics of all the matches in the box. In other words, can we be sure that the results from sampling inspection are consistent with those from complete inspection? Of course, the larger the sample size, the more representative the sample is of the whole lot and so the more accurately all the characteristics of the lot will be reflected in the sample. As the sample size increases it actually becomes the same as the lot.

One of the advantages of sampling inspection is the cost savings produced by the reduction in inspection. However, as you increase the sample size, this advantage will be lost and the cost for inspection is likely to become the same as that for full inspection. Also – and we won't go into details here – there is the very important fact that you cannot carry out complete inspection for products such as matches because this would require all the products to be consumed. In such cases you must do sampling inspection and the sample size must be smaller than that of the lot.

Since it is undesirable to increase the sample size, you need to try to ensure that all the characteristics of the lot are reflected accurately in the sample. Now we know that matches are different from each other, although they look the same. They might be made by workers who have different levels of skill, knowledge, and care toward their work. The products are not always made from the same materials. At a minimum they might be made from different parts of the same pile of material. Also they might be made by different machines that do not have exactly the same characteristics. Therefore, the matches are different from each other. Given these factors, is it possible to determine and classify those differences in order to make sure that the sample is truly representative of the lot?

If one match is chosen from the upper part of the box, then it has a combination of characteristics. If another match is chosen from the bottom of the box, it has another combination of characteristics. However, the result of the sampling inspection doesn't apply to all the matches in the box if matches are chosen only from a particular part of the box. Therefore, we need to choose matches from all locations, such as up, down, near the middle, and near the side of the box, so that we can get a sample that reflects all the characteristics of the lot even if the sample size is small. In this way, each characteristic of the totality of matches is reflected in the sample with a probability determined by its frequency. This is random sampling, and it also provides the answer to the second question.

Even if a sample is chosen randomly, it might not possess some quality characteristics of the lot. It is, of course, a question of probability. Therefore, sometimes it is desirable – even necessary – to take two or more samples, that is, to use double or multiple sampling inspection methods.

The more frequently samples are taken from the lot, the more accurately will the results of the sampling inspection reflect the quality of the lot. This tells us that we get more knowledge for the population by decreasing the sample size and increasing the frequency of sampling rather than by increasing the size and decreasing the frequency. According to the law of probability, the former is the better sampling method. The answer to the question "How many times should we sample from the lot?" depends on the trade-off we make between the required accuracy and the cost of inspection.

Sometimes situations occur in quality control where it is not appropriate to apply random sampling. For example, the exhaust pump for vacuum tubes must be inspected several times a day, because operating the exhaust pump under the same level of efficiency for the whole operating period will definitely determine the final quality of vacuum tubes. To achieve this, every hour we select several vacuum tubes that have finished the exhausting process and measure their extent of vacuum. These tubes are produced at almost the same time, therefore the differences between them are considered very small. However, there may be significant differences between a group of vacuum tubes from one time and a group from another time (not only in individual levels but also in average level). This method of division is called "rational subgroup".

In the example with matches, the question was whether the second match was the same as the first one. Suppose that, although we assumed at first that they were the same, the second one didn't ignite. This suggests that not all the matches are the same. In fact, each match certainly differs from every other match.

From a philosophical point of view, it can be postulated that no two things in the world are the same. All events and things can be similar but there must be *some* differences. You don't get the same value for precision even if you measure the same thing twice with the same aim and methods. This shows that the conditions for those two occasions were not precisely the same.

Statistics accepts this fact and tries to answer the first one of the two probabilistic questions having arisen in relation to sampling inspection by classifying the occurrence of events. In statistical inference the relationship between the events is explored, e.g. the condition of production process and the quality of the produced product; thus it is possible to determine the shape or equation of the relationship at least to a certain extent, if some degree of relationship has been revealed. Moreover, it is also possible to determine the probability of a certain result or event, e.g. of a lot of products being of acceptable level, if other environmental conditions are fixed. The probability of an event is defined as the ratio of the number of desired cases to all possible cases, if all cases are equally likely to occur.

This idea can be explained in terms of a pair of dice. Each die has six sides, so the number of possible combinations made by two dice is $6 \times 6 = 36$.

If 1 occurs on one die, any number from 1 to 6 can arise from the second die. Similarly all other possible combinations can arise for a pair of dice. If each die is a perfect cube, and they are thrown many times and roll over freely, then the combinations displayed by the two dice tends to follow a certain model for the distribution of the occurrence of these events. We show this in the following table.

Phenomenon	Occurrence of phenomenon	Probability
Sum of two outcomes	Frequency	Percentage
1	0	0
2	1	2.78
3	2	5.56
4	3	8.33
5	4	11.11
6	5	13.89
7	6	16.67
8	5	13.89
9	4	11.11
10	3	8.33
11	2	5.56
12	1	2.78

There is only one possible combination producing a total of 2 (which means that each die showed a 1), so the probability of this event after many throws of the pair of dice will be 1/36. If the total is 3, then the probability is 2/36, if the total is 4 then it is 3/36, and so on. In this way the percentage of probability is calculated as in the table.

If the probabilities in the table are plotted and the curve is drawn, you get a distribution curve of frequencies as shown in the graph.

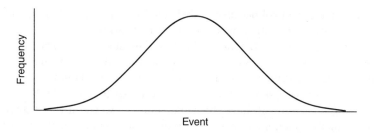

Event

It doesn't matter whether the data are obtained as measurements on a specific characteristic of many individuals or as multiple measurements of an individual. The data will tend to follow this type of distribution for most kinds of measurements on quality or quantity, provided that the measurements are obtained randomly. Of course, the actual variability of the measurements differs each time, but the differences that actually alter the shape of the distribution curve provide much information about the products being inspected. Nevertheless, we can expect that most of the frequency curves will approximate to an ideal distribution, namely the Gaussian or Normal distribution curve. This standard curve shows that most of the values aggregate near the middle of the distribution. In the theoretical model, values are distributed symmetrically and there are few far from the center.

Measures of Central Tendency

The midpoint of the Normal distribution curve, that is to say the central location or central tendency of the measurements, is defined as the "arithmetic mean" of the measurements. This mean is determined by dividing the sum of the measurements by the number of them that you measured. For example, to calculate the mean of five measurements, 6.2, 7.1, 5.5, 8.4, 7.9, calculate their total, which is 35.1. The mean is this total number divided by the number of measurements, i.e. $35.1/5 = 7.02$.

Another measure of central tendency is the "median". If you sort the measurements into increasing order, so that the smallest value is at one end and the largest value is at the other end, then the median is the value exactly in the middle of this sequence. Thus, there is the same number of measurements each side of the median. So if you sort the earlier set of measurements to get 5.5, 6.2, 7.1, 7.9, and 8.4, then the median is 7.1. If an even number of measurements is taken so that there are two central measurements, then the median is defined to be the mean of these two measurements.

The measurement that has the greatest frequency is called the "mode". The mode is the value that occurs most frequently.

The "geometric mean", the "harmonic mean", and the "weighted mean" are also measures of central tendency. However, these are not used very often in statistical quality control because of their complexity of calculation, and they are not as useful as the other measures described earlier.

The arithmetic mean is the most reliable and most accurate measure of central tendency. Therefore, it is used most frequently. Although the median is less accurate than the mean, it is much easier in terms of calculation.

Since the Normal curve is symmetric, the mean (denoted by the symbol $\overline{X}$), median, and mode all have the same value. However, if the distribution of values of the phenomenon under study is not symmetrical, these three measures may have completely different values. In that case, we generally use the arithmetic mean.

When you try to calculate summary statistics with a large number of measurements, it's desirable to use an automatic calculator. This reduces all the subsequent processes to a simple mechanical operation. However, in general, large amounts of data can be handled easily by applying a simple statistical method. What we show you next is the simple and easy way to calculate the mean of a large number of measurements.

When you need to find the mean of 90 carbon resistors, you collect the sample of data by measuring each resistor. The data can then be grouped into small subdivisions called "cells", according to the values of the measurements. You set the number of cells to lie between 10 and 19. In this example, we created the following table by arbitrarily choosing the width of the cell to be 2.5 Ω. (The midpoint of each cell is given by the value of the midpoint between the upper and lower boundaries of the cell. The cell frequency denotes the number of measurements which lie between the lower and upper boundaries of the cell.)

Resistance values of carbon resistors (Ω)

Cell boundaries	Cell midpoint	Frequency	Cell midpoint × frequency
31.25–33.75	32.5	0.0	0.0
33.75–36.25	35.0	4.0	140.0
36.25–38.75	37.5	10.0	375.0
38.75–41.25	40.0	19.0	760.0
41.25–43.75	42.5	23.0	977.5
43.75–46.25	45.0	18.0	810.0
46.25–48.75	47.5	8.0	380.0
48.75–51.25	50.0	3.0	150.0
51.25–53.75	52.5	3.0	157.5
53.75–56.25	55.0	1.0	55.0
56.25–58.75	57.5	1.0	57.5
58.75–61.25	60.0	0.0	0.0
		Total = 90	Total = 3862.5

$$\text{Mean} = \frac{\text{total (cell midpoint} \times \text{frequency})}{\text{total (frequency)}} = \frac{3862.5}{90} = 42.92$$

The mean can be calculated using the values of cell midpoints and cell frequencies as shown in the above table, and it can be calculated more easily based on the values of cell midpoints on an arbitrary simple scale instead on the original scale as below.

Cell midpoint	Frequency	Simple scale	Simple scale × frequency product
32.5	0	–	–
35.0	4	0	0
37.5	10	1	10
40.0	19	2	38
42.5	23	3	69
. . . . 3.166 (see below)			
45.0	18	4	72
47.5	8	5	40
50.0	3	6	18
52.5	3	7	21
55.0	1	8	8
57.5	1	9	9
60.0	0	10	0
	Total = 90		Total = 285

$$\text{Mean on simple scale} = \frac{\text{total (simple scale} \times \text{frequency)}}{\text{total (frequency)}} = \frac{285}{90} = 3.166 = 3\frac{1}{6}$$

In order to find the mean on the scale of the original cell midpoints from the mean on the arbitrary simple scale, you simply align the two scales and note the original scale value that corresponds to the mean on the simple scale. Now it becomes clear that 3.166 on the simple scale falls between 42.5 and 45.0 on the original scale of the cell midpoints. Therefore, the accurate mean is calculated as $42.5 + 0.166 (45.0 - 42.5) = 42.5 + 0.166 \times 2.5 = 42.5 + 0.4166 = 42.92$.

If you compare this value with the mean determined from the previous long calculation, you'll notice that they agree completely. That is to say, if all cells have equal width, no approximation is involved in using the easier calculation based on the simple scale.

The process above can be expressed in the following formula.

(mean on the original scale) = (cell midpoint corresponding to zero on

the simple scale + (cell interval)

× (mean on the simple scale)

When using the arbitrary simple scale, it doesn't matter where you start, provided that you set up the scale in the same direction as that of the original one.

It's the same if you start from integers, such as 1, 7, or −4, etc., instead of starting from zero on the arbitrary simple scale corresponding to the cell's midpoint, 35 on the original scale. In this example the results of calculation will be 42.92, even if you start from any integer. Since the choice of scale is totally arbitrary, if the simple scale is started at a convenient negative value so that zero falls around the middle of the distribution, then the total will be very small and easy to handle so that the calculation can be done by mental arithmetic. However, you should be careful when calculating the sum because of the positive and negative values.

Measure of Variability

The "variability" of a group of data refers to the situation in which measurements are spread out around the central value. The degree of variability is usually measured by the range, the mean deviation, or the standard deviation. Lack of symmetry of variation, in other words, the tendency of measurements to cluster on one particular side of the mean, is measured by skewness. The degree of concentration of the measurements around the median is measured by kurtosis.

The "range" of a group of data (symbol R) is defined as the difference between the maximum and minimum values in the group. To measure the amount of variability, it simply uses just the two values, the maximum and the minimum, without considering the distribution of values in the middle. This is an approximation, but the range is very useful – especially when we have a small number of measurements. However, when we have to perform an accurate analysis or when we need to get precise information, we use another measure of spread.

The "standard deviation" (symbol σ) is a more accurate measure of the spread of data values. It is the square root of the average squared difference between the measurements and their mean value. This is shown by the next equation.

$$\sigma = \sqrt{\frac{(X_1 - \overline{X})^2 + (X_2 - \overline{X})^2 + \ldots + (X_n - \overline{X})^2}{n}}$$

where

$\overline{X}$: mean value of measurements.
$X_1, X_2, \ldots, X_n$: individual measurements
n: number of measurements

The equation above can be easily calculated by one of the methods shown below.

$$\frac{\sqrt{\sum_{i=1}^{n}(X_i - \overline{X})^2}}{\sqrt{n}} = \sqrt{\frac{X_1^2 + X_2^2 + \cdots X_n^2}{n} - \overline{X}^2} = \sqrt{\frac{\sum_{i=1}^{n} X_i^2}{n} - \overline{X}^2}$$

$$\sigma = \sqrt{\frac{\sum_{i=1}^{n} X_i^2}{n} - \left(\frac{\sum X_i}{n}\right)^2} = \sqrt{\frac{n \sum_{i=1}^{n} X_i^2 - \left(\sum X_i\right)^2}{n^2}}$$

$$= \frac{1}{n}\sqrt{n \sum_{i=1}^{n} X_i^2 - \left(\sum X_i\right)^2}$$

The standard deviation corresponds to the turning radius of a system of n particles rotating around an axis, and comes from a technical term in Mechanics. Calculation of σ is much easier using ideas from Mechanics, rather than the above formula.

The term "moment" in statistical analysis has the same meaning as it does in Mechanics. The first moment of a frequency distribution corresponds to the center of gravity of an object. The second moment corresponds to the moment of inertia of a rigid body that has the same mass distribution as the distribution curve of the data. The third and fourth moments are generalizations of the same principle by extension. The meaning of "moment" can be grasped readily by explaining its method of calculation.

The nth moment of a group of data can be determined as follows.

1. Calculate the mean value (divide the sum of the measurements by the number of measurements).
2. Calculate the deviations (subtract the mean from each measurement. You get positive values for measurements that are larger than the mean and negative values for those that are smaller.)
3. Find the nth power of each deviation (n is the degree of the moment you are trying to find).
4. Divide the sum of the nth power of the deviations by the number of measurements. (This is the nth moment of the data. It is also the mean value of the nth power of the deviations.)

The next example shows the calculation of moments.

Suppose we selected 10 springs as a sample from a lot and measured their tensions, resulting in the following data:

43, 41, 43, 36, 42, 45, 43, 41, 42, 44 (measured in grams)

Step 1: When you calculate the sum, you get 42 g. You divide this by the number of measurements, 10, and you get a mean value of 42 g.

Steps 2 and 3: Calculate the first, second, third, and fourth powers of deviations from their mean values.

Measurement	d	Deviation from the mean (d) 2nd power of d	3rd power of d	4th power of d
43	1	1	1	1
41	−1	1	−1	1
43	1	1	1	1
36	−6	36	−216	1296
42	0	0	0	0
45	3	9	27	81
43	1	1	1	1
41	−1	1	−1	1
42	0	0	0	0
44	0	0	0	0
Totals 420	0	54	−180	1398

The first power of the deviation is obtained by subtracting the mean value (= 42) from each measurement.

For example, for the first measurement, the deviation d is defined by

$$d = 43 - \text{mean value}$$
$$= 43 - 42 = 1$$

The second, third, and fourth powers of the deviations are just the square, cube, and fourth power of the deviations. For example, for the fourth measurement,

$d = 36 - \text{mean value} = 36 - 42 = -6$

$d \text{ squared} = (-6)^2 = 36$

$d \text{ cubed} = (-6)^3 = -216$

$\text{Fourth power of } d = (-6)^4 = 1296$

Step 4: Calculation of moments

$$\text{1st order moment} = \frac{\text{total (deviation } d \times \text{frequency)}}{\text{total number of measurements}} = 0/10 = 0$$

$$\text{2nd order moment} = \frac{\text{total (} d \text{ squared} \times \text{frequency)}}{\text{total number of measurements}} = 54/10 = 5.4$$

$$\text{3rd order moment} = \frac{\text{total (} d \text{ cubed} \times \text{frequency)}}{\text{total number of measurements}} = -180/10 = -18.0$$

$$\text{4th order moment} = \frac{\text{total (4th power of } d \times \text{frequency)}}{\text{total number of measurements}} = 1398/10 = 139.8$$

These four moments give us the simplest and most precise information about the shape of the frequency curve compared to any other measures of shape. Several important characteristics related to the distribution of the data are calculated directly by these moments.

The standard deviation can be calculated very easily. It is just the square root of the second-order moment. For the above data, $\sigma = \sqrt{5.4} = 2.32$.

Skewness is denoted by the Greek letter κ, and captures the lack of symmetry of variation. It is calculated by dividing the third-order moment by the third power of the standard deviation. For the above example,

$$\kappa = \frac{-18}{(2.3)^3} = \frac{-18}{12.5} = -1.44$$

Negative skewness indicates that measurements that are less than the mean are located further from the mean value than measurements that are greater than the mean.

Kurtosis is denoted by the Greek letter β_2. This measures the flatness of the frequency curve and is found by dividing the fourth-order moment by the fourth power of the standard deviation. For the above data, the kurtosis is $\beta_2 = \frac{139.8}{(2.3)^4} = \frac{139.8}{29.2} = 4.79$.

When the kurtosis is 3 and the skewness is 0, it means that the frequency curve is symmetrical around the mean value and the curve is the Normal or bell-shaped curve. When the kurtosis is greater than 3, the distribution curve will become more peaked and the data will be more concentrated around the mean value than for the Normal curve. At the same time, some of the data will be further from the mean value than occurs with the Normal curve. If the kurtosis is less than 3, measurements tend to be distributed more uniformly on the measurement scale and there is less tendency for certain values to appear more frequently than other values.

If the skewness is clearly different from 0, kurtosis is easily misinterpreted. Therefore, you have to be very careful in drawing conclusions from the value of the kurtosis when the value of skewness is not small.

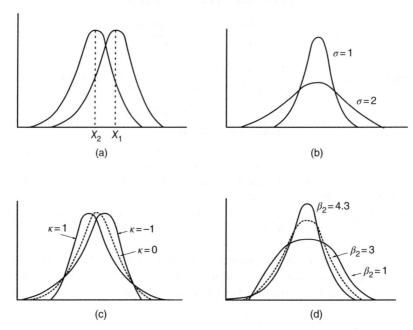

(a) Difference in mean. (b) Difference in standard deviation. (c) Difference in skewness. (d) Difference in kurtosis.

In the earlier discussion of a simple and easy method to calculate mean values, it was explained that the calculation becomes much easier by dividing the data into cells and then calculating using an arbitrary simple scale. We can use the same principle in calculating moments. To reduce the amount of calculation, we may choose an arbitrary working mean and calculate the deviations from this working mean instead of using the true mean. In order to correct for the differences arising using a working mean we can make a small adjustment to the moment. This makes the calculation much easier because we need not calculate the second, third, and fourth power of values with a large number of figures.

The next example assumes that we have selected a sample of 100 light bulbs from the lot, and their electric consumption is measured in milliampere. The results of this inspection are summarized in the following frequency table. We selected 11 cells based on a common cell width of 0.5 mA.

Measurements of electric currents of 100 light bulbs (mA)

Cell boundaries	Midpoint of cell	Frequency
39.25–39.75	39.5	1
39.75–40.25	40.0	1
40.25–40.75	40.5	10
40.75–41.25	41.0	16
41.25–41.75	41.5	16
41.75–42.25	42.0	22
42.25–42.75	42.5	13
42.75–43.25	43.0	11
43.25–44.75	43.5	3
43.75–44.25	44.0	5
44.25–44.75	44.5	2
	Total no. of measurements	100

To calculate the mean value and the standard deviation of these data, proceed as follows.

Calculation of the mean and standard deviation of the data in the above table

Cell midpoint	(f) Frequency	(x) Simple scale	$(f) \times (x)$	(fx^2) $(x) \times (fx)$
39.5	1	-4	-4	16
40.0	1	-3	-3	0
40.5	10	-2	-20	40
41.0	16	-1	-16	16
"A" 41.5	16	0	0	0
42.0	22	1	22	22
42.5	13	2	26	56
43.0	11	3	33	99
43.5	3	4	12	48
44.0	5	5	25	125
44.5	2	6	12	72
	Total (no. of measurements) $= 100$		Total $(+) = 130$	499
			Total $(-) = -43$	
			Grand total $= 87$	

A = arbitrary working mean = midpoint of cell corresponding to zero on simple scale = 41.5 mA.

m = cell interval = 0.5 mA

n = number of measurements = 100

Mean on simple scale =

sum of column (fx)/number of measurements = 87/100 = 0.87

Actual mean mA = value on cell's midpoint scale corresponding to 0.87 on simple scale

$$= A + m \times 0.87 = 41.5 + 5 \times 0.87 = 41.5 + 0.435$$
$$= 41.935 \, \text{mA (actual mean refers to mean on the original scale)}$$

σ^2 (on simple scale) =

(sum of column (fx^2))/((number of measurements) − (square of mean))

$$= \sigma^2 = \frac{499}{100} - 0.87^2 = 4.99 - 0.7569$$
$$= 4.2331$$

Therefore $\sigma = \sqrt{\sigma^2} = \sqrt{4.2331} = 2.06$ (on simple scale).

Therefore σ (on original scale) = $m \times \sigma$ (on simple scale) = $0.5 \times 2.06 = 1.03$ mA

One thing that you have to be careful about when you use a simple scale is that the scale must have the same intervals and increase in the same direction as the cell midpoint. We usually set it to a value that simplifies the calculations. In this case, both positive and negative values appear quite frequently; therefore, you have to be careful not to make a mistake in the calculation.

It is very important to be careful in handling an analysis of the data. We usually treat a large amount of data; therefore, once we misread or overlook a datum, the whole analysis must be carried out again to correct the results. This leads to the loss of time and lowers the efficiency of the department of quality control. Therefore, it is important to handle the data in such a way that the results do not contain any careless mistakes.

Three tables are given below to exemplify a systematic approach to the handling and analysis of the data.

Data table of sampling inspection for quality control

Product _____
Quality characteristic _____
Unit of measurement _____
Specification limits _____ Minimum _____
_____ Maximum _____
Technical Standard no. _____

Production no. _____ Date _____
Production Division No. _____ Inspector _____
Standard production amount per day _____
Sample size _____ per
Inspection No. _____
Signature _____

Item	M'ment [a]	Item	M'ment	Item	M'ment	Item	M'ment	Item	M'ment	Item	M'ment	Item
Sample no.												
Time												
Worker												
Machine												
Total												
Mean $(\bar{X})$												
Max.												
Min.												
Range (R)												

a) Measurement

Simple and easy method to calculate the variation

Calculator _____   Table for calculating moments of the distribution

Checker _____

Example: analysis of measurements of electric currents of 100 light bulbs (mA)

0	1	2	3	4	5	6	7	8
	Cell midpoint	Cell boundaries	Deviation of each cell "A" from origin x	Frequency y	yx	yx^2	yx^3	yx^4
0	39.5	39.25	0	1	0	0	0	0
1	40.0	39.75	1	1	1	1	1	1
2	40.5	40.25	2	10	20	40	80	160
3	41.0	40.75	3	16	48	144	432	1290
4	41.5	41.25	4	16	64	256	1024	4096
5	42.0	41.75	5	22	110	550	2750	12750
6	42.5	42.25	6	13	78	468	2808	16848
7	43.0	42.75	7	11	77	539	3773	26411
8	43.5	43.25	8	3	24	192	1536	12288
9	44.0	43.75	9	5	45	405	3645	32805
10	44.5	44.25 / 44.75	10	2	20	220	2000	20000
11								
12								
13								
14								
15								
16								
17								
18								
19								
20								
			Total =	100	487	2795	18049	127655

m = cell interval = 0.5 mA

$$W_1 = \frac{\Sigma yx}{\Sigma x} = \frac{487}{100} = 4.87$$

$$W_2 = \frac{\Sigma yx^2}{\Sigma x} = \frac{2795}{100} = 27.95$$

$$W_3 = \frac{\Sigma yx^3}{\Sigma x} = \frac{18049}{100} = 180.49$$

$$W_4 = \frac{\Sigma yx^4}{\Sigma x} = \frac{127655}{100} = 1276.55$$

$$V_2 = W_2 - W_1^2 = 27.95 - 23.7169 = 4.2331$$

$$V_3 = W_3 - 3W_1 W_2 + 2\,W_1^3 = 180.49 - 3(4.87)(27.95) + 2(4.87)^3 = 3.1431$$

$$V_4 = W_4 - 4W_1 W_3 + 6\,W_1^2 W_2 - 3W_1^4$$
$$= 1276.55 - 4(4.87)(180.49) + 6(4.87)^2(27.95) - 3(4.87)^4 = 50.4549$$

$$\overline{X} = A + mW_1 = 39.5 + 0.5(4.87) = 41.935$$

$$\sigma = m\,V_1^{1/2} = 0.5\,(4.2331)^{1/2} = 0.5\,\sqrt{4.2331} = 0.5(2.057) = 1.03$$

$$\kappa = \frac{V_3}{V_2^{3/2}} = \frac{3.1431}{(4.2331)^{3/2}} = 0.362$$

$$\beta_2 = \frac{V_4}{V_2^2} = \frac{50.4549}{(4.2331)^2} = 2.8157$$

Simple and easy method for the inspection of variation

Analysis table for measurements obtained by sampling inspection by variables

Material inspected		Page no of code	Date inspected
Production Division	Inspection Division	Inspector's Timecard No.	
Operation of Inspection			
Sampling method No.	Worker's Timecard No.	Lot No.	

Results of Inspection

Lot size	←from here Sample size	Range	First sample Min	Max	Mean	→ to here σ	Second sample Total	Range	Grand total	Range	Variance of lot	Comments

Classification and Analysis of Data

Cell boundaries		Measurements	(f) No. of m'ments	(x)	(fx) (f) × (x)	(fx²) (f) × (fx)
*	*			0		
				1		
				2		
				3		
				4		
				5		
				6		
				7		
				8		
				9		
				10		
				11		
				12		
				13		
				14		
				15		
				16		
				17		
				18		
				19		
				20		
				21		
				22		
				23		
				24		
		Total				

Specification limits   No. of measurements less than lower limit   No. of measurements greater than upper limit

Table for calculation

	Mean ($\bar{X}$)		Standard deviation (σ)	
A	Total of column f =	(f)	Total of column fx^2 =	G
B	Total of column fx =		G divided by A = $\dfrac{G}{A}$ =	H
C	Midpoint of first cell = half the sum of the boundary values Cell interval =		Product of E and E = E^2 =	I
D	B divided by A = $\dfrac{B}{A}$ =		H minus I = H - I =	J
E	Product of E and D = ED =		Square root of J = $\sqrt{J}$ =	K
F	Mean = C plus F = C + F =		Standard Deviation = Product of K and D = KD =	Standard deviation =
Mean	Mean =			

As mentioned before, the ideal shape of frequency distribution is the Normal curve, that is, the Gaussian curve. Its mode, median, and mean are all at the same point, the distribution of the data is symmetrical with respect to the mean, and its skewness and kurtosis are 0 and 3, respectively. The mathematical expression of the Normal frequency curve is shown below.

$$y = \frac{n}{\sigma\sqrt{2\pi}} e^{-\frac{(x-\overline{X})^2}{2\sigma^2}}$$

In this expression y is the value of the vertical coordinate.

n is the number of measurements
$\sqrt{2\pi}$ is 2.5066
σ is the standard deviation of the frequency distribution
e is the base of natural logarithm (2.718)
x is the measurement along the horizontal axis
$\overline{X}$ is the mean of frequency distribution.

Once you calculate the mean $(\overline{X})$ and standard deviation (σ) of a group of data, then just by substituting these values into the expression along with each of the measurements gives you a set of points (x, y) that can be plotted on graph paper as a frequency curve. It will actually help you to understand the distribution of the whole lot or the population. Thus, when you take a sample, in general you can get a curve that has characteristics similar to the curve based on a larger sample or, in other words, the curve drawn according to the result of sampling inspection is expected to show rather accurately the shape of the population.

However, in practice you don't always have to draw an accurate distribution curve. The important thing is not the precise height of the vertical axis but the area under the curve. Therefore, even a free-hand sketch can also help assess the overall distribution.

In situations when you need precise knowledge, you can use a table of the relationship between the vertical coordinate and the cumulative area under the curve. From this table, a "*t*" chart can be prepared, showing the percentage of measurements that fall in the interval from the mean to the sum of the mean and a certain multiple of the standard deviation. In order to get the theoretical frequency, you just multiply this percentage by the total number of measurements.

The next example shows the use of the "*t*" chart.

Measurements of tensile strength of steel (lb/sq. inch) for six samples of steel

141,000	141,500
142,000	141,000
139,500	142,000

First of all, you calculate the mean and standard deviation. In order to make the calculation easier, you should use an arbitrary simple scale.

Calculation of mean and standard deviation of the above data

Cell midpoint		(f) Frequency	(x) Arbitrary scale	(fx) (f) × (x)	(fx^2) (f) × (x)
139,500	1	1	−3	−3	9
140,000		0	−2	0	0
140,500		0	−1	0	0
"A" 141,000	11	2	0	0	0
141,500	1	1	1	1	1
142,000	11	2	2	4	8
	Total	6		Total (+) 5	18
				Total (−) −3	
				Grand total 2	

Area under the normal curve

(x/σ) Distance from the mean, divided by the standard deviation	Ratio of the vertical coordinate for mean $-x$ to that for the mean	Area under the curve less than the mean $-x$ of the normal curve (divided by the total area)
0.0	1.000	0.500
.1	.995	.460
.2	.980	.421
.3	.956	.382
.4	.923	.345
.5	.882	.308
.6	.835	.274
.7	.783	.242
.6	.726	212
.9	.667	.184
1.0	.606	.159
1.1	.546	.136
.2	.487	.115
.3	.430	.097
.4	.375	.081
.5	.325	.067
.6	.278	.055
.7	.236	.045
.6	.198	.036
.9	.184	.029
2.0	.135	.023
2.1	.110	.018
.2	.089	.014
.3	.071	.011
.4	.056	.008
.5	.044	.006
.6	.034	.005
.7	.026	.004
.6	.020	.003
.9	.015	.002
3.0	.011	.001

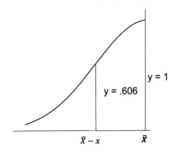

y = 1
y = .606
$\bar{X} - x$ $\bar{X}$

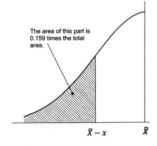

The area of this part is 0.159 times the total area.
$\bar{X} - x$ $\bar{X}$

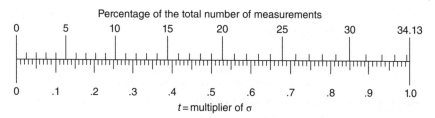

Percentage of the total number of measurements

t = multiplier of σ

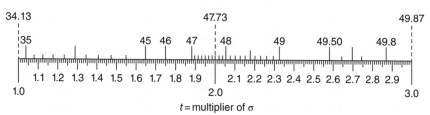

t = multiplier of σ

Note: Percentage of measurements between $\bar{X}$ and $\bar{X}+t\,\sigma$ (or between $\bar{X}$ and $\bar{X}-t\,\sigma$) when they follow a Normal distribution.

A = cell midpoint at "$X = 0$ on simple scale" = 141,000 psi (pounds per square inch)

m = cell interval = 500 psi

n = number of measurements = 6

W_1 = mean (simple scale) = $\frac{\text{sum of raw}(f\ x)}{n}$ = 2/6 = 1/3

$\bar{X}$ = Mean psi = $A + mW_1$ = 141,000 + 500(1/3) = 141,167 psi

σ^2 (simple scale) = $\frac{\text{sum of column}(fx^2)}{n} - W_1^2 = \frac{18}{6} - (1/3)^2 = 3 - 1/9 = 2.89$

σ (simple scale) = $\sqrt{\sigma^2(\text{simple scale})} = \sqrt{2.89} = 1.7$

σ = uncorrected standard deviation in lb psi = $m \times \sigma$ (simple scale) = 500(1.7) = 850 psi

Now we need to explain something about the mean and standard deviation that were calculated by the data. The six measurements of tensile strength were presumed to be a representative sample, but here we also presume that the distribution will be almost the same even if you measure a larger sample. Therefore, we are able to use the sample mean as the midpoint of the Normal distribution. However, you need to use the correction factors or correction coefficients since the standard deviation of the sample is smaller than that of the population.

This correction factor is used only when the sample size is small, and as the sample size increases, there will be no need to use it. You can ignore this factor when the number of measurements is larger than 25. In this case, you make a correction by multiplying the standard deviation calculated from the samples and the factor given by the next table.

Table of correction factors (or coefficients) to estimate the standard deviation of Normal distribution from a small sample.

Sample size	Coefficient	Sample size	Coefficient	Sample size	Coefficient
2	1.77	10	1.08	18	1.04
3	1.38	11	1.08	19	1.04
4	1.25	12	1.07	20	1.04
5	1.19	13	1.06	21	1.04
6	1.15	14	1.06	22	1.04
7	1.13	15	1.05	23	1.03
8	1.11	16	1.05	24	1.03
9	1.09	17	1.05	25	1.03

In the example above,

σ = uncorrected standard deviation = 850 psi
correction factor for sample size 6 = 1.15
$\therefore$ corrected σ = uncorrected σ × correction factor
= 850 × 1.15 = 978 psi

And this leads to the sample distribution curve with mean 141,167 psi and standard deviation 978 psi.

In general, the standard deviation of the population is estimated by the mean of the sample standard deviations adjusted by the correction factor. The mean of the sample standard deviations is equal to the sum of standard deviations of the samples taken from the population divided by the number of samples.

From the above results, you will be able to solve the next questions.

What is the percentage of pieces of steel with tensile strength less than 140,000 lb psi if you take a sample of size 1000 instead of taking a sample of size 6?

What is the percentage of pieces of steel whose tensile strength exceeds 14,300 lb psi?

The "t" value is the value found by dividing the distance from the mean by the standard deviation, in other words it is the value when reading on the scale of the measurement of $\frac{(\bar{X}-x)}{\sigma}$.

Therefore, the value that corresponds to the largest value of 140,000 psi is determined by dividing the difference between this value and the mean by the standard deviation. In other words, 141,167 − 140,000 = 1167; $t = 1167/978 = 1.19$, the value that was measured from the mean to the largest value of 140,000 psi by using the standard deviation as the unit.

From the t-chart, we get 38% as the value that corresponds to 1.19. This means that 38% of all measurements lie between the mean and the mean plus or minus 1.19 × the standard deviation. Since the standard curve is symmetrical, half of all the measurements are less than the mean and the other half are greater. Therefore, the percentage of measurements that are smaller than 140,000 psi will be 50% − 38% = 12%. In other words, if you examine the tensile strength of 100 samples then 12% of the values will be below the aforesaid limit.

In order to know the percentage of values larger than 143,000 psi, it is necessary to determine the t-value that corresponds to 143,000. This is found by using the same method as in previous examples,

143,000 − 141,167 = 1833 psi

$t = 1833/978 = 1.88$

If you refer to the chart, you will notice that 47% of the measurements lie between the mean and the limit of 143,000. Since 50% of all measurements are larger than the mean, the expected percentage of measurements that are larger than 143,000 is 50% − 47% = 3%.

In small samples taken from the Normal distribution, there is a relationship between the range and standard deviation. In other words, it is possible to estimate theoretically the standard deviation by multiplying the range (the difference between max and min) by a certain coefficient that depends on the sample size. It is known that σ estimated by this method is sufficiently accurate for practical use.

Coefficient to estimate the standard deviation from the range.

Sample size	Coefficient	Sample size	Coefficient
2	0.89	9	0.34
3	0.59	10	0.32
4	0.49	11	0.32
5	0.43	12	0.31
6	0.39	13	0.30
7	0.37	14	0.29
8	0.35	15	0.29

When you apply this method to the six tensile strength samples, the range would be the difference between 142,000 and 139,500, which is 2500. The coefficient for a sample of size 6 is 0.39. Therefore, applying this coefficient to the range, you get 975 psi as the estimated value. You will see that the estimated value is very similar to the calculated value by comparing it with 978 psi, which was calculated as the square root of the average sum of squares with the finite-sample-size correction.

Since it is easy to estimate the standard deviation from the range, this is very useful for the analysis of small samples. However, estimated values of the standard deviation using the range are fairly inaccurate and less reliable compared with calculated values, and so, as the sample size increases, they will be relatively more inaccurate. In cases where the sample size is less than or equal to 15, this method is quite useful, but it is not recommended for sample sizes larger than 15.

III

Sampling Inspection

We sometimes classify the measurement of items (such as products and their components) by the method used. One such method is to measure a single item several times and so obtain a more accurate measurement. For example, we may measure the diameter of a metal rod several times.

Usually a single measurement provides an answer to a question about a quantity such as length, width, and number. We can measure the resistance of a coil with prescribed engineering tolerance with just one measurement. However, as mentioned before, the values will be different if you measure the resistance of a coil twice. This is because there are several factors, such as temperature, humidity, individual differences, etc. that may affect the measurement. If the differences caused by such factors are considered significant, then you have to measure the same coil several times in order to get a more accurate result.

Another method is to measure the characteristics for several items of the same type. In this method, one or several measurements are taken for each of several items of the same type, to determine how much they differ from each other. It means measuring the diameter of several metal rods or measuring the resistance of several coils, for example.

In addition to these two methods of measurement, we are also able to classify the methods used to record the results of measurement. For quantitative measurements, the measured values may be recorded as is. For example, when measuring the diameter of a metal rod, the scale value on the micrometer may be recorded in millimeters.

With quantitative measurement, results can also be recorded qualitatively with words such as "It's within the limit" or "It's outside the limit". In this case, we do not record the measured value as is. For example, for the metal rod, we record the measured value as being within the specification limits, or greater than or smaller than these limits, instead of recording the actual value in millimeters.

The properties or characteristics of items are variates that yield different values under different conditions. The resistance of wire is determined by its composition, cross-sectional area, length, and temperature. If one of these

The Road to Quality Control: The Industrial Application of Statistical Quality Control by Homer M. Sarasohn, First Edition. Translated by N.I. Fisher & Y. Tanaka from the original Japanese text published by Kagaku Shinko Sha with a historical perspective by W.H. Woodall and a historical context by N.I. Fisher. © 2019 John Wiley & Sons Ltd. Published 2019 by John Wiley & Sons Ltd.

factors changes, then the resistance of the wire also changes. There are physical laws that give us the relationship between resistance and temperature or other factors. These laws specify the functional relationship between the two variates.

When you cut pieces of wire of the same length from a single bobbin, their resistances will be almost the same, but not precisely the same. The difference is due to the lack of homogeneity in composition, cross-sectional area, length, and other factors. As mentioned above, there are so many factors that may cause the variation, occurring simultaneously, that it's almost impossible to separate each of them. So it is uneconomical to spend time searching for the cause of variation. This is true especially when the degree of variation is small. This kind of variation is called "common (or unassignable) cause variation".

However, when pieces of wire of a given length show significantly different values for resistance, then this difference is considered to have a special factor causing it. Such a large, important difference is referred to as having an "assignable cause" or a "cause to be investigated".

This concept of "common cause" and "cause to be investigated" can be explained by a rifle target diagram. The target on the left shows the bullet marks produced by a good shot.

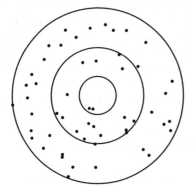 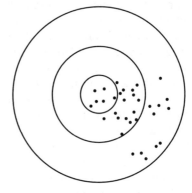

Bullet marks when there are only common causes

Bullet marks when there are both common causes of variation and causes to be investigated

Some bullets have hit the bull's-eye, and others haven't. However, the bullet marks are clustered around the center and scatter in every direction uniformly. It is likely the player cannot explain why some bullets hit the bull's-eye but the others didn't. That is to say, the reason for this kind of variation is caused by chance or "common cause". The distribution of bullets is caused by many factors and the bullets that hit the target successfully are just the lucky ones that were affected favorably by those factors.

In the target on the right, the marks of bullets are clustered on the right-hand side of the bull's-eye. This bias is caused by some specific factor that makes the center of the bullet pattern inconsistent with the center of the target. In this case, from previous experiments we can say that this is due to an assignable cause or a "cause to be investigated", since it is reasonable to expect that the bullets should cluster around the center, if the assignable cause were removed. Bullet patterns that do not cluster in this way indicate that there is a major factor that produces these unforeseen results.

Therefore, the following conclusion can be drawn: if there is no "cause to be investigated" then the bullet marks will be clustered around some central point without showing any bias and if there is no "common cause" then all bullets will hit the same point (which is not necessarily the center) of the target. And when there is neither "common cause" nor "cause to be investigated" at all, then the bullets will all hit the same point in the center of the bull's-eye.

In modern industry, one of the most important problems occurs in mass production, namely, how to produce all products (or manufactured units) with the same quality. However, as we know from the example above, even a good player cannot be expected to hit the same point in the target every time. Similarly, it is impossible to make manufactured items exactly equal. This is why we allow some variability and define limits within which the items will be classified as "good ones". Therefore, our concern is focused on the variables representing the quality and the causes of increasing variability of quality, including both identifiable and unidentifiable variables. Of course, "quality" has to be defined precisely. That is, we have to determine the characteristics to be inspected by expressing the quality in several ways. The importance of this matter was explained in Chapter I, but we shall discuss it again later.

Manufacturers trying to control their production units within defined limits are similar to good players shooting at the center of the target. They have to be careful to ensure that no "cause to be investigated" is introduced into the production process. At the same time, they have to know that sometimes a number of common causes combine to increase the variability to the point that it exceeds the specified limit.

Therefore, by inspecting units during production, manufacturers have to make sure that there is no "cause to be investigated" that could result in unexpected changes, and also make sure that "common causes" remain in their usual state. If a "cause to be investigated" occurs, we need to find it immediately so that we can remove it before it has a deleterious influence on production.

Random sampling is the most convenient method for monitoring the production status systematically. A sample is a set of units chosen as representative of the original population. In this case, "population" refers to the original group of items such as materials, assembled parts, assemblies, and other inspection from which samples were taken. In the case of sampling inspection we sometimes treat a single lot of inspection items as the population, but at other times we regard the lot itself as but one sample from the set of all the continuously produced items. In such cases the population consists of all possible continuously produced items in the production process.

It doesn't matter whether the population is finite or infinite. That is to say, there is no difference if the number of items to be inspected in acceptance inspection is finite or infinite. For example, a lot of raw materials constitutes a finite population. Its size is fixed and the totality of items will be exhausted if you just continue sampling without replacement from the lot. On the other hand, a production process can be thought as an infinite population. It has the potential to provide you with as many lots to be sampled as you need. Any lot can be considered just as a sample when you regard the products of a year or two or more years as the population.

Suppose we have a box containing an almost unlimited supply of cards. Suppose further that each card has a number written on it, and that, if you investigate the distribution of the numbers by taking all the cards out of the box, the frequency curve turns out to be consistent with the Normal curve. Now suppose that we mix these cards well and draw a sample of n cards out of the box at random. The number written on each card is considered as one measurement and is recorded as a measured value. Calculate the mean of the n values (i.e. the sample mean) and return the cards to the box. Mix again well and draw a second sample of size n and calculate the mean value. Repeat this for a third sample, a fourth sample, … , each time calculating the mean for these same-sized samples. Then the plot of a frequency curve for these sample means regarding each sample mean as a single measured value produces a curve similar to the Normal distribution, and you will see that the mean of these sample means will be almost equal to the mean value of the box. However, the standard deviation of the sample means is smaller than that of each individual value on a card. For an infinite population, the standard deviation of the sample means is equal to the value obtained by dividing the standard deviation of individual measured values in the population by the square root of the size of the sample.

In a situation where you take a sample of size n from a finite population of size N, the following result holds:

$$\sigma_{\overline{X}} = \frac{\sigma'}{\sqrt{n}} \sqrt{\frac{N-n}{N-1}}$$

Here, $\sigma_{\overline{X}}$ is the standard deviation of the sample means and σ' is the standard deviation of the population, n is the size of the sample, and N is the size of the population.

However, the mean of all the possible sample means is equal to the population mean, i.e., the mean of all the elements of the population:

$$\overline{\overline{X}} = \overline{X}'$$

These equations hold in the case of a finite population. However, if N becomes much larger than n, then the factor $\sqrt{\frac{N-n}{N-1}}$ will approach 1 and the above two equations approximate to the following equations in the case of an infinite population. Notice that the second equation does not change.

$$\sigma_{\overline{X}} = \frac{\sigma'}{\sqrt{n}}$$

$$\overline{\overline{X}} = \overline{X}'$$

If you take many random samples of size n from a normally distributed population, calculate the standard deviation each time and plot its frequency curve, then it will approximate a Normal curve when the size of sample is more than 25. The mean of this distribution is a little smaller than the population standard deviation. When the sample size is less than 25, the distribution of sample standard deviations will be skewed and its mean will be noticeably smaller than the population standard deviation. Usually, however, even if the sample size is small, we assume that the distribution of sample standard deviations is Normal, and that its mean is consistent with the population standard deviation. In order to make this assumption more reasonable, you simply need to apply the correction factors from the table in Chapter II (on p. 17) to the calculated sample standard deviation:

$$\sigma_{\sigma} = \frac{\sigma'}{\sqrt{2n}}$$

where

$$\sigma' = correction\,factor \times \sigma$$

The equation above is true in the case of an infinite population which has a Normal distribution. If the population is finite or non-Normal, the distribution of any sample standard deviations will be more complicated. However, even in such cases, we usually analyze the data assuming that the measurements follow a Normal distribution approximately but interpret the results of analysis very carefully considering that the results are based on the normality assumption.

If you select two random samples from a single lot of bulbs and measure the consumption current, the two sample means will not generally agree exactly. Also their sample standard deviations are not the same. However, the difference between the means or between the standard deviations will not be significantly large. This is because both samples were taken randomly from a group of bulbs that were made under the same conditions. However, if two samples were taken separately from two different lots, then the difference in mean or standard deviation is likely to be significant unless these two lots were made under the same conditions.

In order to carry out a test for a significant difference, we generally make the sort of assumptions given below. That is, we say the difference is not significant if the difference between the means or between the standard deviations is almost the same as the difference that would occur between random samples from the same population. On the other hand, in cases where the result of the test shows a difference of magnitude rarely observed in random samples from the same population, the difference is considered significant and suggests that there is a cause to be investigated.

Suppose that the mean consumption current of a particular type of bulb is 0.114 A. We suppose this value was calculated from data we collected over several months. Also, suppose that the standard deviation is 0.0034 A. These values are typical of a normal production process and can be considered as the mean and standard deviation of the population. Now we take a sample of 25 bulbs and measure their consumption current and we get a mean of 0.111 A and a standard deviation of 0.0038 A. The question is: "Are these data suggesting that the production process has changed significantly?"

We can answer this question by thinking about how likely we are to get this mean and standard deviation when we take a sample of size 25 from a population that has a mean of 0.114 A and a standard deviation of 0.0034 A.

As mentioned before, the standard error, or the standard deviation of a sample mean based on a random sample of size $n = 25$ is given by

$$\sigma_{\bar{X}} = \frac{\sigma'}{\sqrt{n}}$$

In this case, its estimate is obtained by calculating

$$\sigma_{\bar{X}} = \frac{0.0034}{\sqrt{25}} = \frac{0.0034}{5} = 0.00068$$

The difference between the sample and population means is $0.114 - 0.111 = 0.003$. In order to see the ratio of this difference to the standard error you divide it by 0.00068 to get 4.4 $\left(\frac{0.003}{0.00068} = 4.4 \right)$.

From the t-chart of the Normal curve in Chapter II (on p. 17) the probability of observing a deviation greater in absolute value than $t = 3$ is found by subtracting the area between $t = -3$ and $t = +3$ from 1.

That is,

$$1 - 2(0.4987) = 1 - 0.9974 = 0.0026$$

This shows that the chance of observing a difference greater than $t = 3$ is 0.0026, or approximately 1/400, i.e. once in 400 times. For the sample mean of 0.111, t has the value 4.4; therefore, if a sample is taken randomly from the assumed population with mean value 0.114, the probability that the difference between sample and population means is equal to or greater than $0.114 - 0.111 = 0.003$ is very small, i.e. much smaller than 1/400.

Similarly, it is possible to perform a significance test for the difference between the observed and population standard deviations. You just need to divide the difference between the population standard deviation and the sample standard deviation by the theoretical standard deviation of the sample standard deviation based on the sample of size 25.

That is to say, $0.0038 - 0.0034 = 0.0004$

$$\sigma_\sigma = \frac{\sigma'}{\sqrt{2n}} = \frac{0.0034}{\sqrt{2 \times 25}} = \frac{0.0034}{7.07} = 0.00048$$

$$t = \frac{0.0004}{0.00048} = 0.833$$

According to the t-table the Normal distribution on p. 38 (comment: from this table you can obtain the probability that a standard normal variate takes a value between 0 and abs(t)), the proportion of the area corresponding to $t = 0.833$ is about 0.30. The area greater than $t = 0.833$ on both sides is $1.00 - 2(0.30) = 0.40$. This means that, regardless of the direction, a deviation greater than $t = 0.833$ can occur in random samples from the assumed population once every two or three times ($0.40 = 2/5$). Therefore, we cannot say this difference is significant.

From these data the next conclusion follows. The sample standard deviation is not significantly different from the population standard deviation, that is to say, the variability in the sample does not differ significantly from the variability of the population distribution. However, the sample mean is very different from the population mean, so an important change must have occurred in the production process. They might have used materials with different characteristics. The conditions of the machines might have suddenly changed. Or some other fundamental factors in the production process might have changed. But since there is no significant difference in standard deviation between sample and population, the process itself is considered in control.

In order to check if there is significant difference when comparing the quality of products for each month with the past data collected for a sufficiently lengthy period, you don't usually need to perform a calculation for each sample, as shown above. If the sample size is constant then it is possible to plot the mean and standard deviation of each sample on a graph and draw a control line with the desired significance level. For example, you could draw the line at $t = 2$ or $t = 3$, etc. These lines play the roles of limiting values, and sample means or sample standard deviations that fall outside of these lines are considered significantly different from the past data. This type of graph is called a "control chart" and it has many important uses for industrial engineering, the supervision of production, and general quality control.

The idea above can be used not only for statistical quality control but also for other fields. Let's consider an example.

A telephone relay is an assembly of nine components of equal thickness. When you build up the assembly, the total thickness must be $0.900 \pm 0.006''$, which is the technical standard written on the specification. What should the thickness limit for each component be in order to pass this technical standard? If you want all assemblies of nine components to pass the standard of $0.900 \pm 0.006''$, then (1) each component must be 1/9 of the thickness of the specification, that is $0.100 \pm 0.000\,67''$, or (2) the technical standard of each component must have a tolerance of just over $0.000\,67''$, and when inspecting the newly produced assembled components, omit the ones that are not within the specification limit of $0.900 \pm 0.006''$. However if, instead of using a deterministic limit such as $0.900 \pm 0.006''$ for the standard of thickness of the assembly, the planner uses a statistical representation that shows the mean and standard deviation of the thickness, then it will be possible to show clearly what is required for the components.

According to past experience with the assembly line, most products are clearly within the specification limits, and even if there are some that fall outside the limits, it is known that they are only slightly outside. Then the engineers may perhaps think it is allowable if only a few percent of the products are outside the limits. Let's suppose that, instead of trying to have all products within the specification limits, 5% may fall outside the specification limits of $0.900 \pm 0.006''$. Most of the 5% may be close to the specification limits and there will be fewer products, as the distance from the limit is larger.

By statistical calculation, if the frequency distribution of the thicknesses of assemblies follows a Normal distribution with mean $0.900''$, then from the t-chart of the Normal distribution, 47.5% ($t = 2$) of assemblies should fall between the mean and mean $+2\sigma$ and another 47.5% should fall between the mean and mean -2σ. A total of 95% of assemblies have thicknesses falling within the mean $\pm 2\sigma$ limits. The specification limit has already been set to $0.900 \pm 0.006''$, and since 5% of the assemblies can fall outside this limit, under the aforesaid conditions, we can say, $\pm 0.006'' = \pm 2\sigma$, i.e.

$$\sigma = 0.003''$$

where $0.900''$ can be treated as the mean of the population.

Then it becomes possible to set the specification of each component which is consistent with the specification of the assembly:

σ_i = standard deviation of the thickness of each component
σ_m = standard deviation of the thickness of the sample mean of a sample of size 9 taken from a lot of components, or
= standard error of the mean thickness of a sample of size 9 taken from a lot of components
σ_a = standard deviation of the thickness of an assembly consisting of nine components. This is already set to $0.003''$.

From the above, since the mean thickness of an assembly is nine times the mean thickness of the components used to form the assembly, we can set the standard deviation of thickness of an assembly as nine times the standard deviation of the sample mean of the components of a sample of size nine.

That is,

$$\sigma_a = 9\,\sigma_m$$

The standard deviation of the mean of nine components is $\sqrt{1/9}$ of the standard deviation of each component; therefore, from the equation of standard error,

$$\sigma_m = \frac{\sigma_i}{\sqrt{9}} = \frac{\sigma_i}{3}$$

The relation between σ_i and σ_a is given by eliminating σ_m from the two equations above.

$$\sigma_a = 9\sigma_m = 9 \times \frac{\sigma_i}{3} = 3\sigma_i$$

Since $\sigma_a = 0.003''$ it follows that $\sigma_i = 0.001''$

The above equations means that, if you set the values of mean approximately equal to 0.100″ and of a standard deviation not larger than 0.001″ for each component, the distribution of the assembly consisting of nine components will have mean 0.900″ and standard deviation not larger than 0.003″.

In order to make the meaning of the above discussion clearer, let's discuss it in terms of the probabilities that the assembly and its components fall within specification limits. Thus, according to the laws of probability, at least 95% of thicknesses of these assembled parts fall within the limit $0.900 \pm 0.006''$, if at least 95% of the components are manufactured within the limits $0.100 \pm 0.002''$ and if the assembly has been constructed by taking those parts randomly. The limit $\pm 0.002''$ given to each component is several times wider than the limit of $\pm 0.000\,67''$, which is obtained by another method of dividing the tolerance range of assemblies simply by 9.

Actually, the general problem of how to set specifications for components based on the specification for the assembled products is much more complicated than this example. Not all the components need to have the same thickness, and assemblies need not comprise simple heaps of components. For example, the assembly could be something like a stem fitting into a socket or the electric resistance or inductance of an electric circuit which consists of various resistors or inductors. Also, in such general cases two kinds of explanation can be used as in the above simple cases. The limits for each coil or resistor in an electric circuit, or each component of an assembled product, can be set by applying the basic laws of probability or their extensions. For mechanical, electrical, or any other kind of object, the following equation can be applied as a general formula to any kind of assembly.

$$\text{total } \sigma = \sqrt{\sigma_1^2 + \sigma_2^2 + \sigma_3^2 + \cdots\cdots + \sigma_k^2}$$

In this equation, $\sigma_1, \sigma_2, \sigma_3, \ldots, \sigma_k$ are the standard deviations of the components contained in the assembly and the event of whether or not one part is included in the assembly is independent of the events that other parts are included.

If the chance of an event occurring decreases, its probability also decreases. When the size of the lot is infinite and one of them is defective, then the probability of selecting the defective is $1/\infty = 0$. The probability of an impossible event occurring is zero.

Similarly, as the chance of an event occurring increases, its probability also increases and approaches the upper limit, 1. Probability 1 means that the event must occur.

One basic statistical definition is a definition of probability in terms of a ratio. Thus, the probability of an event occurring is equal to the ratio of the occurrence of the event to the total number of occurrences, assuming that each event has the same chance of occurring. However, in order to think about probability more precisely, you must make sure that each event is independent of each other.

The occurrence of an event is sometimes related to the occurrence of another event. One instance of this is when they are related in the sense of being mutually exclusive. For example, when you throw a die, six possible events might occur. When the die stops rolling, only one surface is on the top. There is no possibility of two faces being on top at the same time.

On the other hand, when the occurrence of an event is unrelated to the occurrence of another event, we say they are independent of each other. Throwing two dice at the same time can produce two independent events. The reason for this is that the upward face of one die does not have any effect on which face of the other die is upward.

If there is a set of events only one of which can occur (there is no possibility of two or more events occurring at the same time), then the probability of the set of these events occurring is the sum of the probabilities of occurrence of each event belonging to the set. For example, in the case of two dice, the probability of a total of 2 is $1/36$, of three is $2/36$, of four is $3/36$, and of five is $4/36$. Therefore, when two dice are thrown, the probability of getting a sum of 4 or less is $6/36$ or $1/6$. The probability of getting 12 or less from two dice is $36/36$, or 1, so this event definitely occurs.

The probability of several independent events occurring is the product of the probabilities of occurrence of each event. For example, the probability of a coin showing a particular face and a die showing 1 at the same time is $1/2 \times 1/6 = 1/12$.

When you think about this formula for calculating the probability of an event, you need to count the number of all possible occurrences and the number of occurrences favoring the event of interest. Therefore, it is necessary to investigate ways of counting the number of possible occurrences of events and the number of ways to arrange things.

In sampling problems with a population size of some thousands and a sample size of some hundreds, the number of possible cases is huge, but it is quite easy to solve if you use the theory of permutations and combinations.

A permutation of several objects refers to different ways of arranging them. The possible permutations when choosing two letters from a set of three letters *a*, *b*, and *c* are *ab*, *ac*, *ba*, *bc*, *ca*, and *cb*. The possible permutations when choosing three letters from the same set are *abc*, *acb*, *bac*, *bca*, *cab*, and *cba*. The number of permutations when choosing *m* objects from *n* can be thought about as follows. The first object chosen can be any of them. The second one is any one of the other *n* − 1. The third one is anyone of *n* − 2. By the same reasoning, repeating this method up to the *m*th step, the *m*th object may be any of *n* − *m* + 1. Every permutation is included in this; therefore, the number of permutations when choosing *m* from *n* is given by their product, i.e.

$$P_m^n = n(n - 1) \cdots \cdots (n - m + 1)$$

If all *n* objects are chosen at the same time, *m* will be equal to *n* and the equation becomes as shown next.

$$P_n^n = n(n - 1) \cdots \cdots 1 = n!$$

(*n*! is the symbol for the factorial function and means the product of all integers from 1 to *n*.)

The combination of several things means the different ways of choosing them without taking account of the order of arrangement. Combinations when choosing two from three letters *a*, *b*, and *c* are *ab*, *ac*, and *bc*. The letters *ab* and *ba* are different permutations but the same combination.

Let's think about combinations for *m* objects. The number of permutations of *m* objects is *m* factorial. That is to say, *m*! permutations can be produced from a combination of *m* objects. This gives us the relationship between the number of combinations and the number of permutations.

$$\text{permutations} = m! \times \text{combinations}$$

That is

$$P_m^n = m! \times C_m^n$$

C_m^n is the symbol denoting the number of combinations when choosing *m* from *n*. This symbol is often shown as $\binom{n}{m}$.

From the previous discussion, the number of permutations is as follows

$$\text{Permutation}(n, m) = n(n - 1) \cdots \cdots (n - m + 1)$$
$$= m! \times C_m^n$$

Therefore,

$$m! \times C_m^n = n(n - 1) \cdots \cdots (n - m + 1)$$
$$C_m^n = \frac{n(n - 1) \cdots \cdots (n - m + 1)}{m!}$$

Multiplying the numerator and denominator by $(n - m)(n - m - 1) \cdots$ $\cdots 1 = (n - m)!$ yields

$$C_m^n = \frac{n(n-1)\cdots\cdots(n-m+1)(n-m)(n-m-1)\cdots\cdots1}{m!(n-m)(n-m-1)\cdots\cdots1}$$

Therefore,

$$C_m^n = \frac{n!}{m!(n-m)!} = \binom{n}{m}$$

Based on the above mathematical guide for statistical inference, let's consider some problems related to random sampling.

Specifying an Acceptable Proportion of Defective Items

Following the preliminary five steps, i.e. Step 1 to Step 5, explained in the first chapter, in order to establish a quality control system in an industrial company, it is necessary in Step 6 to set a limit on the proportion of defective items allowable in a lot of products that passes inspection. This means that you should perform sampling inspection in such a way as to make sure that the proportion of defective items (or sometimes number of defects) in the lot does not exceed the allowable maximum.

The easiest way to ensure this is to perform sampling inspection for every lot. There may be a lot in which the number of defective items reaches the limit. Then you should do the following. For a given sample size, if the number of defective items in the sample does not exceed some specified acceptance limit, you accept the lot; otherwise, reject it. The acceptance limit of the number of defective items in the sample can be calculated for a given sample size so as to ensure that the probability of accepting a lot whose proportion of defective items is greater than or equal to the specified acceptable level is less than some specified value. This value is called the "consumer's risk".

In terms of probabilities, this problem can be stated as follows. If you are given the size of a lot, the upper limit for the proportion of defective items in the lot to be accepted, and the probability of accepting the lot with the proportion of defective items equal to this upper limit, then you are able to determine the sample size and acceptable defect level in the sample which satisfy these conditions. This is the common explanation of the problem of sampling inspection. In other words, to simplify things, the problem is to determine the probability of finding m defective items in a sample of size n taken from a lot of size N containing M defective items.

When you take out an item at random from a lot of N items, the probability that it is defective is given by the ratio M/N. However, as you continue to take out items one by one without replacement, the probability of taking out a defective item will change and become much more complicated. That is because the ratio of the number of defective items to the total number of items will change in the sampling process.

When the first item is taken out, the ratio among the rest will be $(M - 1)/(N - 1)$ if the first item is non-defective and $M/(N - 1)$ if the first item is defective. After two items are taken out, this ratio will be more complicated, and when a large number of items are taken out, calculating the ratio will be almost impossible. That is why we need to develop a general probability formula that can be used for sampling inspection.

Let us use the theory of permutations and combinations. The number of ways selecting n items from a lot of size N is expressed by $\binom{N}{n}$. Here we do not take account of the order of sampling the items.

The number of ways of selecting m defective items is $\binom{M}{m}$, because these items are taken out from a total of M defective items contained in the lot.

In the same way, the number of ways of choosing $(n - m)$ non-defective items from a total of $(N - M)$ is $\binom{N-M}{n-m}$.

The number of ways of selecting m defective items and $(n - m)$ non-defective items at the same time is clearly equal to the product of $\binom{M}{m}$ and $\binom{N-M}{n-m}$. This gives the number of cases in which m defective items are contained when drawing a sample of size n from a lot of size N containing M defective items. This yields the formula

$$P_{n,m}^{N,M} = \frac{\binom{M}{m}\binom{N-M}{n-m}}{\binom{N}{n}}$$

In order to show how this general formula is used, let's calculate the probability of not getting any defective items when a sample of size 2 items is taken randomly from a lot of size 20 containing 8 defective items and 12 non-defective ones.

In this case, $N = 20$, $M = 8$, $n = 2$, $m = 0$

Then, using the formula shown above,

$$P = \frac{\binom{8}{0}\binom{20-8}{2-0}}{\binom{20}{2}}$$

$$= \frac{\binom{8}{0}\binom{12}{2}}{\binom{20}{2}}$$

$$= \frac{\dfrac{8!}{0!(8-0)!} \quad \dfrac{12!}{0!(12-2)!}}{\dfrac{20!}{2!(20-2)!}}$$

$$= \frac{8!}{0!8!} \times \frac{12!}{2!10!} \times \frac{2!18!}{20!}$$

Therefore,

$$P = \frac{12 \times 11}{20 \times 19} = \frac{132}{380} = 0.347$$

(Note that $0! = 1$.)

This general formula is exact. However, it is annoying to calculate factorials. There are tables giving the factorials up to $N = 1000$, but even with such tables the calculation takes a lot of time and requires hard work. Therefore, it is desirable to find an easier way of doing it.

When you look back and see how the general formula was derived, you will notice that it is the change of the ratio of the number of defective items to the total number of items in the sampling process that makes the calculation complicated. If this ratio doesn't change, then the calculation will be simpler.

In order to fix the ratio of defectives to total, the size of both the total and the number of defectives needs to be very large. For example, if the lot has 1,000,000 items and the number of defective items is 10,000, the ratio doesn't change very much even if a large number of items are taken out as a sample.

Therefore, if we can assume the following three conditions, there will be no problem in regarding the ratio of defectives to total as constant.

1. The size of the lot is very large.
2. The number of defectives in the lot is large.
3. The sample size is very small compared to the lot size (usually less than 1/10).

In order to derive a simple formula, note that the probability of choosing a defective item in a single trial of sampling is equal to the proportion of defective items in the lot just before this single trial. Thus, the probability of taking out one defective item in the first single trial is M/N, which is equal to the fraction defective in the lot. The terminology "proportion of defectives" and "fraction defective" will be used interchangeably.

We assume here that the lot is large enough and that the ratio M/N can be regarded as constant during the sampling process. The fraction defective, that is to say, the probability of choosing a defective item when taking out one item, is constant. We denote this by the letter p. The proportion of non-defective items, that is to say the probability of choosing a non-defective item, is also constant. We denote this by the letter q. It is obvious that, $p + q = 1$ or $p = 1 - q$.

The probability of the first item being non-defective is q. The probability of the next item being non-defective is also q. Therefore, the probability of both being non-defective is given by the product of the two probabilities, $q \times q = q^2$. The probability of the third item being non-defective is also q. The probability of the first three items being non-defective is q^3. Reasoning this way, the probability of the first n items being non-defective is q^n. If all n items are non-defective then the number of defects is zero. In other words, the probability of not having any defect in a sample of n items is $P_0 = q^n$.

The probability of the first item being defective and the next three items being non-defective is $p \times q \times q \times q = pq^3$. Therefore, the probability of getting a defective item first and then getting $n - 1$ non-defective ones is pq^{n-1}. In typical inspection, the order of defective and non-defective items is not considered a problem.

There are n possible positions for a defective item, namely, the first, the second, the third, ... , nth and, since all have equal probability of occurrence, the probability of getting one defect in a sample of n items is $P_1 = npq^{n-1}$, when you don't take account of ordering.

In the same way,

$$P_2 = \frac{n(n-1)}{2!}p^2q^{n-2}$$

$$P_3 = \frac{n(n-1)(n-2)}{3!}p^3q^{n-3}$$

$$P_4 = \frac{n(n-1)(n-2)(n-3)}{4!}p^3q^{n-3}$$

$$P_m = \frac{n(n-1)(n-2)\cdots\cdots(n-m+1)}{m!}p^mq^{n-m}$$

$$= \binom{n}{m}p^mq^{n-m}$$

Now we show the results once again:

$$P_0 = q^n$$

$$P_1 = npq^{n-1}$$

$$P_2 = \frac{n(n-1)}{2!}p^2q^{n-2}$$

$$P_m = \binom{n}{m}p^mq^{n-m}$$

The probabilities are consistent with the terms of a binomial expansion. That is,

$$(a+b)^n = a^n + na^{n-1}b + \frac{n(n-1)}{2!}a^{n-2}b^2 + \cdots\cdots$$

$$+ \binom{n}{m}a^{n-m}b^m + \cdots\cdots + b^n$$

We can make the substitution $a = q$, $b = p$ in this formula. And to simplify things, we can make the substitutions shown below:

$$P_0 = q^n$$

$$P_1 = n\frac{p}{q} \times P_0$$

$$P_2 = \frac{(n-1)}{2}\frac{p}{q} \times P_1$$

$$P_3 = \frac{(n-2)}{3}\frac{p}{q} \times P_2$$

This is a convenient set of equations when you need to calculate each probability using the binomial expansion of $(q + p)^n$. With these equations you can calculate immediately the probability of 0, 1, 2, ... defective items in a sample of size n. Now we show how to use these binomial calculations with the data from the previous example.

$$N = 20$$

$$M = 8$$

$$P = M/N = 8/20 = 0.40$$

$$q = 1 - p = 1 - 0.40 = 0.60$$

$$n = 2$$

$$m = 0$$

$$\therefore \ P_{n,m}^{N,M} = \binom{n}{m} p^m q^{n-m} = \binom{2}{0} p^0 q^{n-0} = \frac{2!}{0!(2-0)!}p^0 q^{n-0} = \frac{2!}{2!}p^0 q^{n-0}$$

Here we eliminate the common factors of denominator and numerator, and by substituting $p^0 = 1$ we get

$$P_{n,m}^{N,M} = q^n = (0.6)^2 = 0.6 \times 0.6 = 0.36$$

Therefore, the probability of getting two non-defective items and no defective items from a sample of size $n = 2$ is 0.36. This result is a few larger than the true value, 0.347, which was calculated previously. The Binomial equation gives the precise and correct result when the population is infinite (when the lot is considered very large). However, as in this example, when the sample size is less than 1/10 of the size of the lot the true value and the value from the Binomial distribution are very close to each other.

Up to here we have only treated quantities obtained by counting, such as the proportion of defective items, but now we are going to treat quantitative characteristics which can change continuously. When we measure the plate currents of vacuum tubes which are operating under the same conditions, then the results will show values which are slightly different from each other within a certain range. For example,

$$0.2637, 0.2636, 0.2702, 0.2598 \ldots$$

Even if there are vacuum tubes which show the same value, we will be able to find several tiny differences if we measure them more precisely.

These quantities are the quantitative characteristics that can change continuously. However, in quality control we sometimes treat them qualitatively. In that case, we only investigate whether the parts, assembled parts, or products meet the required quality specification by measuring them. Acceptance or rejection doesn't depend on how closely they conform to the specification but is determined solely by whether they meet the requirement. For example, if the specification of the plate current of a vacuum tube is set to more than 270 mA and less than 300 mA, then only one vacuum tube will be accepted.

In these cases, it suffices for instruments that are used in factory inspections to simply have a special indicator board that only indicates the range of acceptance. The indicator board with ordinary scale marks is not only not needed but also easily misinterpreted. The actual instrument reading is not needed: we only want to know whether it is consistent with the specification or not. Such an indicator board separating conforming items and defectives should be colored (for example, a yellow part showing the acceptable range, and red on each side of the yellow to show the ranges for rejection) so that inspection becomes much more correct and effective. By doing this the inspector can easily tell whether the product is conforming or defective just by looking at the position of the instrument needle.

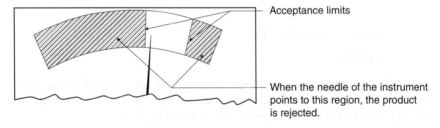

Acceptance limits

When the needle of the instrument points to this region, the product is rejected.

These methods can be applied to the cases of other instruments and to products other than vacuum tubes.

Sampling inspection is often applied in the form of sampling inspection by attribute. Products that don't meet the quality specification are regarded as defective items and the quality level of the total lot is indicated by the percentage of defective items, termed the "fraction defective" or "percent defective". We denote the fraction defective by the symbol p.

As we showed in the previous example, it is possible to create a frequency distribution for the sample fraction defective. Its standard deviation is given by the following formula:

$$\sigma = \sqrt{\frac{p(1-p)}{n}}$$

As an example, let's think of a large lot of screws. This lot contains 25% defective items. When you take a sample of 200 screws, how many defective items will be found in the sample? Using the above formulae, it is expected that, if you take samples many times, 68% of samples will contain 44 – 56 defects (21.9 – 28.1%) and 95% of samples will contain 36 – 62 defects (15.8 – 34.2%). These numbers are determined by multiplying the size of the sample and the percentage of the defects.

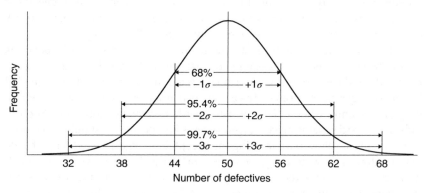

Each percentage indicates the ratio of the area under the curve corresponding to limits shown in the figure to the total area under the curve.

$$c = pn = 0.25 \times 200 = 50$$

$$\sigma = 6\ (3.1\%)$$

If we draw a frequency curve of this screw example, its curve will approximate the bell-shaped Gaussian distribution, but for exact calculations it is better to use a method based on the binomial expansion of $(a + b)^n$.

In the above example, if we set the percentage of defectives to 1% instead of 25%, then the most frequent number of defects that appears with samples of size 200 is $0.01 \times 200 = 2$.

If we substitute $a = 0.99 = q$, $b = 0.01 = p$, $n = 200$, the binomial expression will be

$$(0.99 + 0.01)^{200}.$$

When this expression is expanded, the first four terms will be

$$0.1343 + 0.2713 + 0.2726 + 0.1817 + \dots$$

These values are the probabilities of containing 0, 1, 2, and 3 defectives, respectively, in a sample of size 200.

It is hard work to expand a binomial expression. There is an approximate method that is easier than this, using the Poisson distribution:

$$\frac{1}{e^c} + c\frac{1}{e^c} + \left(\frac{c^2}{2!} \times \frac{1}{e^c} \right) + \left(\frac{c^3}{3!} \times \frac{1}{e^c} \right) + \cdots + \left(\frac{c^n}{n!} \times \frac{1}{e^c} \right)$$

where

e = the base of the natural logarithm (2.718 28)
p = the acceptable percentage of defects (lot tolerance percent defective)

Therefore
$c = np$ = the maximum acceptable number of defectives per n (= 200) items in the lot.

If we apply the Poisson approximation to the example, the first four terms will be:

$0.1353 + 0.2707 + 0.2727 + 0.1804$

The terms show the probability of finding 0, 1, 2, and 3 defects respectively in a sample of size 200.

The values obtained by using binomial expansion and those by using Poisson approximation agree very well with each other. We may choose a method with appropriate accuracy required for the practical application; therefore, there will be no problem to use a simple method like Poisson approximation depending on the properties of particular sampling inspections.

We have spent many pages on the theory of statistics and probability, so we should make some remarks to avoid misunderstanding. Some readers may think that high-level complicated mathematics occupies most or all of statistical quality control. There are some scientists who placed unreasonable emphasis just on the theoretical side of quality control. It is an undesirable attitude. If this attitude prevails, then there's no hope of applying quality control to get good results. Foremen and other managers might be afraid of this high-level mathematics and business people might wonder what use scientific problems have in their world. You should not overestimate the role of the theory of statistical inference in a quality control system. Quality control is one kind of application of mathematics but, fundamentally, it's not an area for theoreticians to work in. Practical application is the point to be emphasized.

There are many people who think the mathematics of probability is a very difficult science that cannot be understood using commonsense. However, in the definition of the theory of probability, which states it is an area of applied mathematics dealing with the effect of chance, there's no secret at all. When you think about a simple event, everybody can understand it easily. For example, suppose we have a box with 5000 glass balls with 5% of them black and 95% white. When you take out 100 glass balls randomly as a sample, then the number of black balls can be any number between 0 and 100, but if you take the average of sampling experiments over a long time then it will be expected to be 5.

Thinking this way, the chance of there being 0 black balls in a sample of size 100 is about 1 in 165; and the chance of there being 10 or more is once in 40. Using the Poisson approximation, it will be easy to estimate the probability for any combination of black and white balls.

Now suppose that there are 5000 glass balls but we don't know the number of black balls. Using the Poisson formula, if we draw a sample of size 25 and suppose that the proportion of black balls is 5%, then 30% of the samples should contain 0 or 1 black ball, 3 or more black balls will occur once in eight times, and finding 4 should occur once in 40 times.

If we assume the proportion of blacks in the lot is 8%, finding 0 or 1 in the sample occurs once in eight times, 3 or more once in three, and an occurrence of 5 has a chance of once in 40. Then we examine the number of black balls in the sample from the lot. If there are many cases of having 4 black balls, then the proportion of blacks in the lot is more than 5%. On the other hand, if there are only a few cases (for example once in 40 times) of having 5 black balls, it means that the proportion of blacks in the lot is not more than 8%. Therefore, we set 8% as the percentage of defects and establish the method of sampling inspection and continue until we realize that we should correct this supposition.

We have been talking about statistics and probability, so now let's think about their application to quality control in industry based on this explanation.

IV

Application of Sampling Inspection

As an example of the basic applications of sampling inspection, we shall explain a method that is used to judge the quality of the lot of products or materials. In contrast to the traditional, conservative approach to inspection of checking every unit of the lot, sampling inspection only requires looking at a portion of the given lot. In other words, sampling inspection is the new concept in competition with 100% inspection. The 100% inspection method is a way of screening products and it has been used widely. Products are inspected to check whether they have the required characteristics, and separated into acceptable, unacceptable, or defective items. Acceptable items are shipped to customers and defective items are returned to suppliers and either modified or rejected. (From here on, the words "consumers" and "producers" are each used in a broad sense, so we need to be careful. In other words, consumers are not only customers of products but may also users of intermediate products or of materials in the production process. Correspondingly, producers are not only factories but also can be suppliers of materials or intermediate products produced in the processes preceding the production process.)

The shortcoming of 100% inspection is its high cost because it requires a large amount of inspection.

1. It is comparatively expensive.
2. It is hard and boring work for inspectors.
3. It requires many inspection instruments and facilities.
4. It cannot be used as a guide for investigating the current conditions of production since the production process will have been completed at the point of inspection.
5. Its accuracy is not always high.

100% inspection isn't always 100% accurate. It can even produce more variable results than sampling inspection because of the attitude or interest of the inspectors. For example, in one factory small metal parts were being polished, and 8% defective items were eliminated as a result of 100% inspection. However, when the accepted items were re-inspected, they were found

The Road to Quality Control: The Industrial Application of Statistical Quality Control by Homer M. Sarasohn, First Edition. Translated by N.I. Fisher & Y. Tanaka from the original Japanese text published by Kagaku Shinko Sha with a historical perspective by W.H. Woodall and a historical context by N.I. Fisher. © 2019 John Wiley & Sons Ltd. Published 2019 by John Wiley & Sons Ltd.

to contain a further 2% defective items. And at another factory, Bakelite parts were subjected to 100% inspection and 5% were eliminated. However, the next day 0.7% defective items were found and on the third and fourth days, inspectors found 1.1 and 0.72% defective items, respectively. The third example to show the imperfect nature of 100% inspection is a factory making components for resistors. They used 100% inspection and eliminated 5% defective items. However, they found 1% defective items in the remaining lot.

There are many more similar examples. Thus, there are situations where sampling inspection is more appropriate than 100% inspection. The cost is lower, and the time required for the inspection is shorter. Theoretically, 100% inspection has the highest accuracy, but actually not necessarily so, as explained in the above paragraph. Taking into account inspection errors sampling inspection has sometimes higher or much higher accuracy.

In implementing sampling inspection, the first thing to do is to determine the method of sampling products randomly. Sampling should be done so as to avoid bias. That is to say, all characteristics of the lot should be reflected in the sample. Any item among acceptable items, defective items, and items with any other characteristics of the lot should be contained in the sample with equal probability.

We should be very careful that the cyclic changes of production processes do not bias the characteristics of the sample. We should avoid sampling with exactly the same time intervals. The reason is that samples will not be properly representative if producers know when sampling will take place and change the way they work accordingly.

The second stage of implementing sampling inspection is to determine which of two basic approaches in current engineering use to apply. The first approach is aimed at assuring lot quality, and the second is aimed at assuring average outgoing quality.

To assure lot quality, the first thing to do is to select an upper limit for the lot percentage defective and then to design the inspection process in such a way that only a small prescribed percentage of lots that contain in excess of this percentage defective are accepted. For example, it may be planned that for lots containing 5% defective items no more than 10% are accepted (1 lot in 10). Of course, there is only a small probability of accepting a lot with 10% defectives. However, this approach to sampling inspection doesn't mean we are certain that all those lots with 10% defectives will be rejected. It just means the probability of this type of event occurring is very small. This probability can be calculated.

In this type of application, the average quality of shipped (accepted) lots is higher than the specified critical percentage of defective items. According to the sampling protocol described above, there is only 1 chance in 10 of acceptance, even if many lots containing the critical percentage of defective items are presented. Lots that are assessed as unacceptable will be returned to the producers or screened using 100% inspection. However, this should be applied only for rejected lots and there is no need to do 100% inspection for all lots. This is one of the merits of sampling inspection from the perspective of economy.

The average quality of outgoing lots using the above sampling inspection is a function of the average quality of the inspected lots (average process quality) and the lot size.

When we apply quality assurance to a lot, it is usually the case that one or other of the following two methods is used. These are single and double sampling inspection methods. (There are more complicated methods but we shall not go into detail here.)

Single sampling inspection is carried out as follows.

1. Draw a sample randomly from the lot.
2. Inspect the items in the sample and record the number of defective items.
3. Compare the number of defectives in the sample with the acceptable number of defectives provided beforehand, and judge whether to accept or to reject the lot according to the result.
4. If the lot is rejected, identify the defective items in the lot with 100% inspection and improve them or exchange them with acceptable ones. Sometimes the lot is returned to the producer.

Double sampling inspection has the psychological effect of giving us hope that there is still a chance for the lot to be accepted. It is carried out as follows.

1. Draw the initial sample randomly from the lot.
2. Inspect the initial sample and record the number of defects.
3. According to the number of defects in the first sample, the lot will be accepted, rejected, or sent to the second sampling stage.
4. Draw the second sample randomly from the lot.
5. The sum of the number of defects in the first and second samples is compared with the acceptable number of defectives provided beforehand and the acceptability of the lot judged according to the result.
6. Rejected lots are sent for 100% inspection to identify the defective items, and the defective items are modified or exchanged with acceptable items. Sometimes, the lot is returned to the producer.

The second method for assuring quality is intended to determine the limit of average outgoing quality. In this case, a standard of acceptance is specified for the average quality of the lots after completing inspection. That is to say, the quality of the lots after inspection changes depending on the average quality of the lots before inspection (average process quality), the size of the lot, and the specified limit of average outgoing quality level. Single inspection, double inspection, or other sampling inspection methods can be used also for assuring average outgoing quality according to requirements.

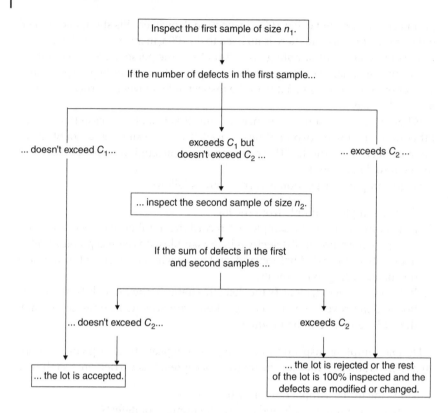

n_1: the number of items in the first sample.

n_2: the number of items in the second sample.

C_1: the acceptable number of defective items in the first sample.

C_2: acceptable number for the sum of the number of defective items in the first and second samples.

Summarizing these facts, a sampling inspection plan to assure the lot quality comprises sampling inspection of each lot and the resulting treatment for each lot (accepted, rejected, or re-inspected). In order to make a decision, we compare the percentage of defects found in samples with the pre-defined acceptable percentage of defective items. Then the accepted lot is assured not to contain more than the acceptable number of defectives.

On the other hand, the sampling inspection plan to assure the average outgoing quality does not assure the quality of each lot; it only provides assurance that the average quality of many lots will be within the regulated standard. Therefore, some accepted lots will contain more than the acceptable percentage of defective. Since these are offset by lots with low percentages of defective items, the overall percentage of defects in accepted lots will be lower than the pre-defined percentage. Taking these into consideration, statistical tables have been created for the standard of acceptance in this type of sampling inspection. That is to say, if lots are inspected and their results are compared with the guidelines, and if the average percentage of defects does not exceed the acceptable level, then all the lots are classified as acceptable. However, if the average percentage of a group of lots exceeds the standard of regulation, then we must adjust the average quality toward the acceptable level by eliminating lots that fail the standard. Those eliminated lots will be rejected or sent to be re-inspected.

When we try to apply these sampling inspection plans in practice in a factory, we cannot always say which approach is better. The second approach to assuring average outgoing quality is probably more economical, whereas the first approach enables us to control the production process better. It is especially effective when the production rate is not very high. Thus, each approach has both advantages and disadvantages, but both of them are very useful and can give us almost the same degree of quality assurance.

It is not theoretically possible to determine which approach is more suitable in any given situation; rather, this should be determined by studying the particular conditions for each case. Here are some examples of such conditions.

When customers purchase the product in large quantities on an ongoing basis, you can provide them with satisfaction by assuring the average quality of the products that the customers receive. On the other hand, if the buyers purchase several lots only occasionally, then an approach that assures the quality of each lot is more suitable.

In the case where certain parts or products are going to be used in the next process, we should sometimes use an assurance plan for lot quality that accepts lots for which the percentage of defective items is larger than a critical limit for the next process. However, if in the long run the data indicate that each lot is at the acceptance level, then we should apply an assurance plan for average outgoing quality, to save on inspection costs.

In assuring lot quality, the lots which have the critical acceptance percentage of defectives are accepted only at a certain rate – normally 10%. That is why, in general, the quality of accepted lots is better than the actual standard. However, this is only true when average lot quality or average process quality is better than the critical acceptable percentage of defects in the lots. Where this doesn't hold, the number of defects found in sampling inspection will increase and, therefore, the cost will become higher.

The second type of sampling inspection plan assures the average quality of the outgoing lots; therefore, it is economical when the average quality of the process is approximately equal to the standard for the average outgoing quality level. However, regardless of which type is adopted, the basic principle for determining the inspection plan in any given condition is to choose the cheapest and easiest plan among those that can assure the required quality.

There is no way to decide automatically which of 0.25, 1, or 2% to use to determine the acceptable percentage of defective items in a lot or to determine the standard of average outgoing quality. We can only determine it by careful study of the particular conditions of a given factory. Particular conditions include such things as: the capability of the process, the skill of the mechanics, engineering efficiency, market requirements, and other factors that affect production and sales. Management experience, experience with processing, data collected during past production, the history of the company, the cost of achieving the required quality level, and management requirements for quality assurance are all factors that should be considered in determining the standard of quality.

Cost is an important factor to consider when you handle this problem, since inspection is related to economic problems. When the size of the sample is the smallest and the number of the items to be re-inspected is also the smallest, it is obvious that the inspection cost is minimized. However, as mentioned before, since the accuracy of inspection goes down as you reduce the size of the sample, you sometimes have to re-inspect lots that have been rejected in order to confirm their real quality. We need to find a compromise between the two conflicting requirements in order to assure the maximum quality with the minimum cost. The diagram below shows the relationship between cost and sample size.

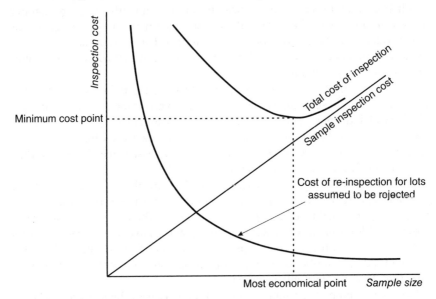

Here we want to provide assurance that the lot quality is 1% defective items. That is to say, we want to determine the minimum sample size required for the lot tolerance percent defective to be 0.01 for the lot. The consumer's risk, that is, the probability of lot with this percentage or higher percentage of defects, is normally set to 10%. Suppose that the sample size is 100. The critical number of defective items allowed per n items in the lot ($c\bar{c}$) is $0.01 \times 100 = 1$. Substituting these values in the formula utilizing the Poisson distribution and taking the first five terms gives us the following:

$$0.368 + 0.368 + 0.184 + 0.061 + 0.015$$

The terms in this summation show, respectively, the probabilities of a sample of size 100 containing 0, 1, 2, 3, and 4 defects.

Therefore, if we set the critical number of defects in a sample to 0 (for convenience, set $\bar{c} = 0$ here) as the standard for a lot to be accepted, then 37% of the lots that contain 1% defective items will be accepted. However, we have already put in place a standard that states that no more than 10% of lots with 1% of defectives should be accepted. Therefore, we now know that a sample size of 100 is not sufficient to ensure the required level of quality.

In order to find the sample size that meets this requirement, you just have to do the same calculation for samples of sizes 200 and 300. If you calculate values for these two cases using the Poisson distribution, then you will find that in the former case 14% of lots with 1% defectives will be accepted; and in the latter case, 5% will be accepted. From this calculation, it is clear that the minimum sample size to assure the required level of quality lies between 200 and 300. With further calculation, you will find that the sample size should be 230.

To summarize the above: take a sample of 230 items randomly from a lot, and if you don't find any defectives in the sample then the lot is accepted. In this way, the quality of the accepted lot is assured to have less than 1% defective items 9 times out of 10.

In order to assure this level of quality, the sample size needs to be at least 230, but there are lots of possible combinations of sample size and acceptable number of defects to assure the same level of quality.

Given the size of the sample (n) and the lot tolerance percent defective (p), the acceptable number of defectives per n items in the lot is $c = p \times n$. Various combinations of n and p such that c is equal to 2.30 and the first term of the formula using the Poisson distribution is equal to 0.10 (consumer's risk) will work as sampling inspection plans to assure the same level of quality with the case obtained above, that is the minimum sample size is $n = 230$ given that $\bar{c} = 0$ and consumer's risk $= 0.10$.

For example,

c	=	p	×	n
2.30		0.01		230
2.30		0.02		115
2.34		0.03		78
2.32		0.04		58
2.30		0.05		46

When you plot these data as in the following chart with the cases $\bar{c} = 1, \ldots,$ 4, you'll be able to read the acceptable number of defectives $(\bar{c})$ in a sample directly from the curves. The vertical axis of this graph is the lot tolerance percent defective or the acceptable percentage of defective items in a lot, and the horizontal axis is the sample size. One of these curves corresponds to the case $\bar{c} = 0$. If the number of defectives in the sample is set to zero, then the sample size can be read from the curve of $\bar{c} = 0$, given the lot tolerance percent defective p (with the consumer's risk equal to 10%).

In the above example, the first term in the formula based on the Poisson distribution is the probability that the number of defective items in the sample is 0 when the percent of defective items in the lot is equal to 1%. The sum of the first two terms is the probability that the sample contains no more than one defective item. Therefore, if you choose a $\bar{c}$ value consistent with the sum of first two terms being 0.10, you can get information about sample size n and lot tolerance percent defective p for inspection plans with a consumer's risk of 0.10, when the acceptance number of the defectives in the sample is set to 1.

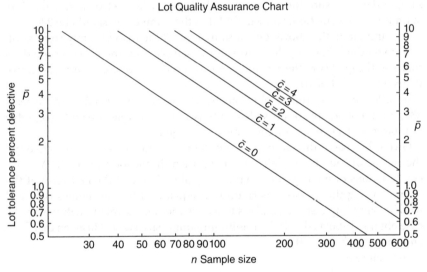

Chart to determine the sample size (n) and the acceptable number of defective items in a sample ($\bar{c}$) for the case where consumer's risk is equal to 0.10.

Given this lot tolerance percent defective p, there are many possible combinations of the sample size (n) and the acceptable number of defective items in the sample ($\bar{c}$), so you can choose the sample size to match the given acceptable number of defective items. In the same way, for $\bar{c} = 0, 1, 2, 3$, it is possible to read the sample size appropriate for the required level of quality directly. For example, if the lot tolerance percent defective is 3%, all of the combinations assure the same degree of quality:

When $\bar{c}$ is 0, n is 77
When $\bar{c}$ is 1, n is 130
When $\bar{c}$ is 2, n is 178
When $\bar{c}$ is 3, n is 222
When $\bar{c}$ is 4, n is 226

Normally, it is better to set the sample size sufficiently large that the acceptable number of defective items is 1 or 2, until it becomes clear that it is safe even if you reduce the size of the sample.

With sampling inspection, there is the possibility of rejecting a lot that should have been accepted. This is the opposite of consumer's risk and is called "producer's risk". Therefore, the cost of sampling inspection consists of two factors. One is the cost of inspecting samples and the other one is the cost of re-inspecting the rejected lots that should have been accepted. Suppose that the quality of the lot is very good; in other words, there are very few defective items in the lot. If you set the lot tolerance percent defective to 1% and the consumer's risk to 10%, then the minimum sample size for which the acceptable number of defective items in the sample is 0 is 230, as explained before.

Now suppose that the average quality of the process or average percent defective of the lot to be inspected is 0.35% and that the size of the lot is 5000. The Poisson distribution calculated for 230 items sampled from a lot with percentage of 1% defective items is shown next.

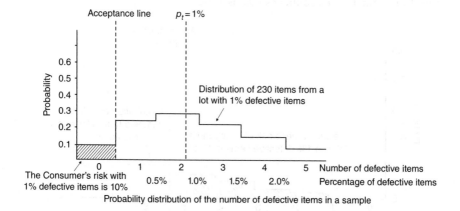

Probability distribution of the number of defective items in a sample

Next, we show the calculation for the case in which the average process quality is 0.35%. In this case, about 55% of the lot is classified to be rejected.

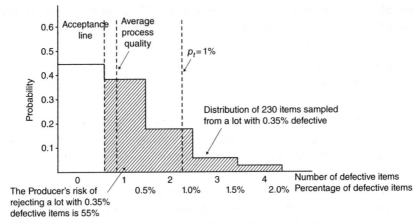

The Producer's risk of rejecting a lot with 0.35% defective items is 55%

Probability distribution of the number of defective items in a sample

Therefore, when the sample size is 230, an average of $(5000 - 230) \times 0.55 = 2623$ rejected items are to be re-inspected. In this case, the average inspection number is 2853. Thinking about many lots that have to be inspected in this application of assurance of lot quality, you should be taking account of the fact that more than 50% of items are to be re-inspected.

Now let's set the sample size to 925. The following diagrams show the Poisson probabilities obtained by taking two samples, one from a lot with 1% defective items and the other from a lot with 0.35% defective items. In these diagrams it is shown that the lot is not rejected even if the sample contains five defective items and that in the case of a lot with 0.35% defective items the probability of rejection, namely, the producer's risk, is 11%.

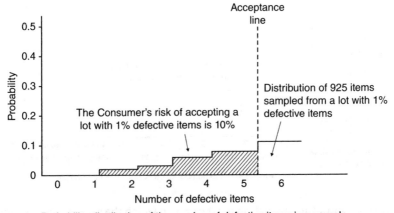

Probability distribution of the number of defective items in a sample

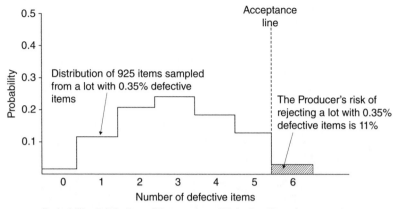

Probability distribution of the number of defective items in a sample

For a lot with 0.35% defective items the average number of items to be inspected is 1373, which is found by adding the sample size, 925 and $(5000 - 925) \times 0.11 = 448$, the number of rejected items to be re-inspected. You will notice that this is half the earlier number obtained in the case of sample size 230. Therefore, even if the cost of sampling inspection increases, there is a decrease in re-inspection cost and so the total cost of the latter becomes much smaller.

In the same way, the table of inspection standards is created by specifying the process average quality, lot acceptable percentage of defective items and other conditions, and by calculating the most economical size of the samples. Several tables relating to the assurance of lot quality and the assurance of average outgoing quality for single or double sampling are shown at the end of the final chapter. Using these tables, you may easily determine the suitable size of the samples. If you know the average process quality of the lot, then by determining the acceptable quality or the standard of average outgoing quality, you will be able to choose the most economical sample size for a given lot size. And also you will be able to find the maximum acceptable number of defective items appropriate to each circumstance.

V

Quality Control of the Production Process

Well-managed factories go through a certain process when they start producing a new product. First of all, expert engineers carefully design and check whether the design is suitable for the expected use by examining it thoroughly. They also check whether it is commercially viable, whether the manufacturing arrangement can be set up easily, whether the required method of manufacturing conforms to the technical standards of the company and doesn't need any special machines or techniques, and whether the assembly time is not too long. Then the engineering department prepares drawings, lists, and specifications of materials for the product, sets of working instructions, and other documents required by the manufacturing department, and ensures that they contain no defect and are accurate.

The manufacturing department needs to have a good understanding of its role in the overall production process by studying the documents and information prepared by the engineering department. Manufacturing engineers are required to adapt the facilities of the factories and develop the skills of the workers in order to manufacture the new products by taking into account the reason for designing new products. Manufacturing engineers have to prepare process specifications on the basis of the quality standards determined by design engineers. Each manufacturing process that has to be carried out by the workers in each step of manufacturing must have a sheet of clear instructions. The instruction sheet shows them what to do, how to do it, when to do it, which instruments should be used, and where to send the output of the process.

Manufacturing engineers have to prepare the standard of inspection based on the reason for designing the new products. On standard documents, they have to identify clearly the quality aspects in relation to materials, purchasing parts, process products, and final products. The issue of quality starts with the arrival of materials. Quality is maintained by suitable workers, suitable methods, and suitable machines. From the moment the materials arrive, to the shipping of products, there's no error of quality that can be permitted in any stage of the process.

The Road to Quality Control: The Industrial Application of Statistical Quality Control by Homer M. Sarasohn, First Edition. Translated by N.I. Fisher & Y. Tanaka from the original Japanese text published by Kagaku Shinko Sha with a historical perspective by W.H. Woodall and a historical context by N.I. Fisher.

The purpose of statistical quality control is to put in place the methodology described above. It is planned so that no one in the company is indifferent to maintaining quality. It provides a means for management to recover quickly when the manufacturing process to maintain the good quality falls into disorder.

This methodology is necessary for the development of companies. In order to manage efficiently and effectively it is important to conduct the manufacturing operation in a manner that follows the specification designed by engineers. It is important not only for avoiding accumulation of defects but also to prevent losing the trust of the customer. Variation in product necessarily causes high cost.

Therefore, in order to operate more efficiently and effectively it is sometimes necessary to change the method of manufacturing. However, when there's not enough evidence to justify the need for change, it's not unusual for managers to refuse to change the manufacturing process. It used to be difficult to establish this since there was no quantitative evidence. However, these days, not only can statistical quality control provide such evidence but it can also indicate precisely where changes are especially required.

The purpose of this approach is to prevent recurrence of out-of-specification production; in other words, quality control is used to reduce the variation in product quality. Sampling inspection is useful not only for checking the quality of the manufactured products but also for checking on quality when accepting manufacturing materials. Moreover, sampling inspection is used for quality control of all the manufacturing processes. However, in order to apply this control effectively, the basic idea and its importance should be generally understood by everyone involved.

The purpose of statistical quality control of a manufacturing process is not to get information about the quality of each product but to know immediately when the quality starts to diverge from the quality standard. This is necessary to ensure that the manufacturing operation is kept in control. In other words, it is to ensure that the quality of most products falls within specification limits. If the quality is within specification limits, then we can expect that the products manufactured in this period of time will be accepted with a certain high probability. However, if they diverge from specification, that is, if the result of sampling inspection shows they fall outside the specification limits, then it is necessary to find and eradicate the cause of this change so that production returns to the state of statistical control.

Statistical methods, especially the quality control chart, are designed to record the manufacturing process continuously and to provide immediate notification of the occurrence of the out-of-control state, so that we can take action to search for and eliminate the causes. We shall explain the control chart later on.

There are a number of methods for detecting assignable causes in manufacturing process. One of them is as follows: plot the standard frequency distribution from the result of sampling inspection carried out during a period in which most of the products conformed to specification. Then, by conducting sampling inspection continuously, the results can be displayed as frequency distributions. By comparing the current frequency curve with the standard curve, we can see whether the current quality is the same as, better than, or worse than the quality for the control period.

In the same way, it is possible to compare the sample standard deviations (σ_σ) between the current and control periods. If this difference is large, then it means that current products have greater variation than for the control period. In this case, some modification is required.

It is not just the manufacturing department but also the engineering department which benefits from the application of statistical quality control. We can immediately ascertain whether the manufacturing method is suitable or not for the specification required by the design, by comparing the control limits and information obtained for the manufactured products. If many of the products are out of specification limits, then you need to decide, based on the information obtained by quality control and the technical knowledge of the manufacturing process, whether you should change the method of manufacture or change the specification limits by adjusting the design to suit the capability of the factory. The information gathered from quality control clearly shows where things need to be improved, for example, by much more careful production, or by teaching the operators to carry out the standard procedures more precisely, or by replacing the tools by newer or better ones.

However, before establishing the system of manufacturing process control, it is necessary to perform a technical investigation to ascertain what kind of manufacturing method is suitable as well as whether the materials and the processes are appropriate for producing the products with required quality, and whether technical requirements are consistent with manufacturing capability. Any action or setting of control limits should be done after carrying out this investigation. And the change of quality characteristics within the control limits should not be regarded as a change due to assignable causes. However, when quality characteristics range outside the limits, then we should think that there are some assignable causes relating to this change.

In manufacturing companies, it is desirable that, rather than people with appropriate skills being distributed amongst various organizations, they be grouped together in the department of quality control. These people should be able to report directly to the chief operating officer, who has the power to control quality and cost. Such people will be directors, production managers, or manufacturing managers, depending on the nature of the company.

When you establish the department of quality control, you should not initially allocate routine factory inspections that do not relate directly to quality control to quality control engineers. However, it is desirable that all the inspection functions are under the control of the same chief. Once the department of quality control is established, it should select the methods of quality control, be the technical leader of sampling inspections, calculate the limits for acceptance inspections, create the control charts, prepare the technical standards for sampling inspections, and start leading the investigations required to establish the manufacturing methods to make products using processes that are in control. As a part of these duties, the quality control engineers should choose several important quality characteristics of the products in consultation with the design engineers. These characteristics should collectively capture good performance of the products.

They may be characteristics that can be measured only at the final product stage, or possibly they can be measured both during the production process and at the final inspection. For example, when you apply this to the manufacture of vacuum tubes, the current of the tube of an electric amplifier is a characteristic that should be measured by a quality control engineer during the process, but it should be done also after the final factory examination. On the other hand, the gap between the cathode and the grid should be measured only during the production process.

After choosing the quality characteristics to be measured and determining the quality standards and the specification limits, the next job of the newly established department of quality control should be to analyze the production process. This is done in order to identify which types of causes are likely to produce defects at the various steps in the production process. First of all, investigate the relation between the selected quality characteristics and the technical standard or instruction sheet for production. Then investigate the effect that each manufacturing process has on the quality characteristic, especially the effect of the material or intermediate product from the previous step in the process, the machine operation, or the skill of the operator.

Next, look at the method of inspecting a given characteristic in the factory. Here, it is necessary to pay attention to factors that cause error in measurement. Sometimes variation in quality derives not from manufacturing errors but from inspection errors. Poor practices in inspection arise from using an inappropriate measuring instrument, a poorly functioning instrument used for examination, bad habits of the inspector, or from errors in the inspection standard given to the inspector.

Once these investigations have been completed, we have to determine the unit quantity of products. That is, we need to fix the size of the lot of products. Is it the total production per day, per machine per day, or per hour? This should be determined by taking account of the convenience of handling the products and the required productivity. In any event, the size of the lot should be clearly defined. And once you have determined it, then you should not readily change it unless you have good reasons to do so.

Determining the size of the lot is very important, as shown next. A certain factory had six machines to make vacuum tubes, and a lot was defined as the total production of the six machines per day. The standard specified that the probability of accepting lot with more than 2% defective items should not be larger than 10%. However, since the actual percent defective of this production process was about 3%, many lots were judged to be rejected, and the increased inspection cost due to the need for re-inspecting rejected lots caused significant financial loss.

This problem was solved by changing the definition of lot size from total production per day to production per machine per day. The situation altered immediately and it became clear that five machines were working almost perfectly but one machine was producing 18% defects. Therefore, it emerged that, prior to changing the lot size, most of the rejected lots were being produced solely by this defective machine, and that after changing the definition, almost all the lots produced by the other five machines would be accepted by the inspection. Thus it became possible to focus on repairing this machine.

Once you have determined the size of the lot, the next step in building the quality control system is to determine at which stage of the manufacturing process to begin the inspection or examination of quality characteristics. This is because the aim of quality control differs from that of traditional 100% inspection. In the case of traditional 100% inspection, which involves separating conforming items from defective items, the production operation for the products finishes when the factory's inspection has finished. We do nothing if any defects are found. If defects are produced, then we repair them, remake them, or just throw them away. Anyway, they represent a loss for the company.

However, the aim of the quality control is to find those parts of the process with a tendency to produce defects during manufacturing and to prevent the production of defects. Therefore, the department of quality control has to measure the level of quality at a very early stage in the manufacturing process. We cannot wait until the whole process has finished.

The next problem in establishing the quality control system is to plan how to summarize and classify the information obtained by the sampling inspections. The standard inspection chart shown later will help to solve these problems since there is a fixed minimum size of sample for the given size of the lot. However, with a sufficient amount of experience in quality control, it might be possible to find another suitable sampling method apart from the ones shown in the chart.

Moreover, it is necessary to establish methods that specify how, why, where, and by whom samples are to be taken from the lot. Also, how and where to treat and analyze the samples has to be determined. This treatment will be done by the department of quality control separately from current factory inspection. The department of quality control will use both the data from factory inspections and from the sampling inspections for quality control. Therefore, the department of quality control will require other inspection instruments and inspection places. Of course, the quality control inspector has to be trained in how to record and classify the data collected on quality characteristics. The above information has to be written clearly in the technical specification and the inspectors have to know how to follow the standard correctly.

Now we need to decide whether we record measured values as continuous data for each inspected item or just write "accepted" or "rejected" following the specification requirements for attribute data. This decision determines which control chart is to be used: a control chart for the average ($\overline{X}$), the standard deviation (σ), the percentage of defects (p), the range (R), or some other form of control.

The recording method for sampling inspection is also important. Results of inspection have to be saved so that they are easily matched with production time or other events occurring during the manufacturing process. By doing this, it becomes easier to find the cause of a change in the sequence of measurements. The form of control chart which records the quality control inspection data has already been mentioned on p. 31.

Based on prior investigation of the manufacturing process, we have to decide how to group or classify the results of inspection of selected samples and corresponding manufacturing conditions. The important things relating to this classification differ depending on the sampling plan used. Which sampling method to use for each quality control application should be determined by taking account of the specified aim, the differences in cost, compatibility with the capabilities of the manufacturing department, relationship with the quality control department, and the required relative accuracy.

Let us summarize the above. Careful consideration of the situation of selecting small groups sensibly makes us realize some common important elements. If you take small groups separately without considering the manufacturing conditions, there will be some questions or some differences, such as the difference in production time or difference in materials and machines. If you try to characterize the population, you have to depend on random sampling. However, there are two points to note about this.

1. When taking samples at equally spaced time points, the cycle should not be synchronized with the cycle of any production condition which may influence quality.

2. Selection of samples should not be done using a fixed timetable. For example, if you take samples every hour, workers who are aware of this will make allowances, and that influences the quality of products.

If the samples are taken systematically, so that they are all the same size, then the application of quality control will be much easier and useful. Several samples are taken once an hour or once a day. These are selected to be the same size or the same proportion of the lot. It is also important to take samples in such a way that they reflect well the current conditions of the manufacturing process. In determining the size and frequency of sampling it should be taken into account that, in general, taking small-sized samples frequently provides a better reflection of the actual situation than taking larger samples less frequently.

Here we have to clarify the term "control chart", which has been used many times so far. As mentioned later, a control chart is similar to a street. A street has a pavement and a strip on each side. Normally, there is gutter at the edge of the pavement. The width of the pavement is usually determined by the traffic conditions. The road used only by bicycles can be narrow but a wider road is required for cars. A control chart is the same as this. The width of the central safety zone is normally determined by the characteristics of the manufacturing method of the factory. The control limit is similar to the edge of the pavement. This is outside the safety zone.

In the case of a street, it is clear that the driver of a car will notice as soon as its wheels stray from the pavement. He will turn the steering wheel, step on the brake, or take some other action required; otherwise, a perilous situation might soon occur. It is the same with a control chart. The results from quality control inspection are plotted on a control chart. If the plotted points are approaching a control limit or warning limit, it is necessary to take action to prevent any serious problems.

Moreover, if the car is trapped in the gutter the driver will be totally confused. It is the same for the factory. If the product enters the danger zone (which corresponds to the gutter for a control chart), there must be a problem in its quality. In such cases it is necessary to stop production until the causes for the out-of-control state have been removed or a new method of operation has been introduced so that the control state is achieved.

If all causes found to have an effect on quality have been eliminated from the manufacturing process and if all plotted points lie within the control limits, then the quality characteristics can be said to be "in control". However, if the process is brought into control and stays there for a long time, then the manufacturing method cannot produce higher quality than that. If you want higher quality, you cannot achieve it without making a fundamental change to the manufacturing method.

Ditch	Danger	Upper specification limit
Bank	Warning	U.C.L.
Paved road	Safe	Average quality
Bank	Warning	L.C.L.
Ditch	Danger	Lower specification limit

There are two methods to determine the control limits. The first is to calculate them based on past manufacturing experience. Collect manufacturing records from the past several months and calculate the quality characteristic quantities using the statistical method described in Chapter II. This is to determine the quality limits for future production based on production up to the present. The base period of time should correspond to normal operations. The measurements are divided into groups based on production time and other conditions and divided further into smaller subgroups based on reasonable criteria. Calculate the average of each subgroup. Now calculate the average of all the measurements (or a sufficiently large subset of measurements to be representative of the whole). This is sample grand mean, denoted $\overline{\overline{X}}$. Then calculate the range or standard deviation from the same set of measurements.

Based on these results, a frequency distribution curve can be drawn by plotting the measured values along the horizontal axis and the frequency of occurrence of each characteristic value on the vertical axis. In the case of a Normal distribution, the curve will exhibit symmetry around the average value. Regardless of the shape of the curve, the general properties of the distribution can be ascertained by studying the spread of values. Once the skewness, kurtosis, and standard deviation have been calculated, the distribution curve can be explained by the aforesaid method.

For a symmetric distribution, you can approximate the value of the standard deviation by finding the average deviation of the observed value from the arithmetic mean. Alternatively, you can obtain an approximate value for the standard deviation graphically. Draw a perpendicular from the top of the symmetric distribution curve to the horizontal axis, then draw a line parallel to the horizontal axis at the 3/5 point from the bottom of the perpendicular and find the intersection P with the distribution curve. The horizontal distance from the perpendicular to point P gives an approximation to the standard deviation.

However, for most situations in which the control chart is constructed for the first time, one would expect to use the accurately determined value of the standard deviation which was calculated by the equation in Chapter II.

Now we find the control limits of the control chart by using the overall mean and range or estimated standard deviation. These control limits are found by adding and subtracting three times the standard deviation to the mean. The coefficient "3" in formula $\overline{\overline{X}} \pm 3\sigma$ is not determined by probability theory, but it is the appropriate expression based on historical experience with production. That is, this 3-sigma control chart is useful in defining the limits for taking action, since it is expected that 99.7% of products of a well-controlled production process will be contained within the limits.

In this case, since the engineering department does not prescribe the standard, the method is based on historical factory information. Therefore, this limit describes the capability of the manufacturing process assuming that the materials, the data collecting period, and the production method do not change.

However, in deciding to use $\overline{\overline{X}} \pm 3\sigma$ as the control limits for quality characteristics, you need to bear in mind the design specification. For example, let's say that the mean value is 7 and the standard deviation is 0.50; then $\overline{\overline{X}} \pm 3\sigma = 7 \pm 1.50$. Therefore, the upper limit is 8.50, and lower limit is 5.50. When you think of design specification, is it acceptable to use control limits ranging from $\overline{\overline{X}} + 3\sigma$ to $\overline{\overline{X}} - 3\sigma$ (8.5 to 5.5)? In other words, can we use product that falls in this range? Might there be some incompatibility resulting from using components exhibiting such a spread? Does this fit the design specification? Don't we need to shrink the limits closer together? These problems should be examined and resolved by the engineering department. This is a quality issue relating to practical use of the product, and it needs to be checked whether something may go wrong using product manufactured within this range of variation.

If we need to narrow the limits, we might set them to $\overline{\overline{X}} \pm 2\sigma$ or $\overline{\overline{X}} \pm \sigma$. With the current manufacturing method, the percentage of conforming products will be 95% or 68% respectively, and the number of defects will be huge. Therefore, if there is a need to narrow the limits without increasing the probability of producing defective items, there will need to be a fundamental change in the manufacturing process aimed at reducing the standard deviation to a small value.

The second way to determine the control limits is to use a formula based on statistical theory to handle the quality control problem. This is used widely and it is clearly more accurate because it is derived theoretically.

If you divide the past measurements into more than 25 subgroups of equal size, then the central line of the control chart of mean values is the mean of the subgroup means. The control limits are calculated by the formula $\overline{\overline{X}} \pm A_1 \overline{\sigma}$ (where A_1 is a coefficient that varies depending on the subgroup or sample size n, and $\overline{\sigma}$ is the mean value of subgroup standard deviations). This can be expressed as

$$\overline{\overline{X}} \pm 3 \frac{\overline{\sigma}}{c_2 \sqrt{n}}$$

Here, c_2 is a coefficient relating to the standard deviation that depends on the subgroup or sample size. The next table contains coefficients used to calculate the control limits for samples sizes from 2 to 25. For sample sizes greater than 25, you can calculate the coefficients from the table by extrapolation. However, for large-sample calculations the amount of error is not so significant even if you neglect this modification.

The table of coefficients for control charts is composed of three parts, namely, coefficients relating to the control chart for the mean, coefficients relating to the standard deviation control chart, and coefficients relating to the range control chart. The coefficient A_1 is given in the second column of the first part and c_2 is in the first column of the standard deviation control chart.

In the control chart for the standard deviation, the central line is given by the mean of the standard deviations of all subgroups or samples, and the control limits are determined by calculating $B_3 \times \overline{\sigma}$ and $B_4 \times \overline{\sigma}$. Here B_3 and B_4 change depending on the size of the subgroup or sample sizes and are found from the table. The limit is also found from the following calculation:

$$\overline{\sigma} \pm \frac{3\overline{\sigma}}{c_2 \sqrt{2n}}$$

It is possible to create a control chart if you are able to calculate the control limits by any one of the methods described above. Let the vertical coordinate represent the value of subgroup or sample mean or standard deviation, and the horizontal coordinate represent manufacturing time or subgroup number. Plot the central value and control limits obtained from the past measurements on the vertical axis, and draw three lines parallel to the horizontal axis through these points. These three lines form the control chart. Now you start to plot the sample means or standard deviations obtained in the production process on this control chart. Points are plotted based on the scale of the vertical axis and the numbers along the horizontal axis. If production is in control, then all the points fall within the two control limit lines. If a point falls outside the area between the two control limit lines, it suggests that something has caused a change in quality. However, if this happens rarely, and if the objective is to create an initial control chart, you can just eliminate such points and calculate the control limits using the remaining points.

Table of coefficients for control charts

Type of control chart	Control chart for mean			Control chart for standard deviation					For small samples				
	Coefficients for control limits			Coefficient for central line	Coefficients for control limits				Coefficient for central line	Control chart for range			
										Coefficients for control limits			
Sample size n_1	A	A_1	A_2	c_2	B_1	B_2	B_3	B_4	d_2	D_1	D_2	D_3	D_4
2······	2.121	3.759	1.880	0.5642	0	2.064	0	3.658	1.128	0	3.686	0	3.268
3······	1.732	2.394	1.023	0.7236	0	1.948	0	2.692	1.693	0	4.358	0	2.574
4······	1.500	1.880	0.729	0.7979	0	1.859	0	2.330	2.059	0	4.698	0	2.282
5······	1.342	1.596	0.577	0.8407	0	1.789	0	2.128	2.326	0	4.918	0	2.114
6······	1.225	1.410	0.483	0.8686	0.003	1.735	0.003	1.997	2.534	0	5.078	0	2.004
7······	1.134	1.277	0.419	0.8882	0.086	1.690	0.097	1.903	2.704	0.205	5.203	0.076	1.924
8······	1.061	1.175	0.373	0.9027	0.153	1.653	0.169	1.831	2.847	0.387	5.307	0.136	1.864
9······	1.000	1.094	0.337	0.9139	0.207	1.621	0.227	1.774	2.970	0.546	5.394	0.184	1.816
10······	0.949	1.028	0.308	0.9228	0.252	1.594	0.273	1.727	3.078	0.687	5.469	0.223	1.777
11······	0.905	0.973	0.285	0.9300	0.292	1.570	0.312	1.688	3.173	0.812	5.534	0.259	1.744
12······	0.866	0.925	0.266	0.9359	0.324	1.548	0.346	1.654	3.258	0.925	5.593	0.284	1.717
13······	0.832	0.884	0.249	0.9410	0.353	1.528	0.375	1.625	3.336	1.026	5.646	0.308	1.692
14······	0.802	0.848	0.235	0.9453	0.378	1.512	0.400	1.599	3.407	1.121	5.693	0.329	1.671
15······	0.775	0.817	0.223	0.9490	0.401	1.497	0.423	1.577	3.472	1.207	5.737	0.348	1.652
16······	0.750	0.788	·····	0.9523	0.422	1.483	0.443	1.557	·····	·····	·····	·····	·····
17······	0.728	0.762	·····	0.9551	0.441	1.470	0.462	1.539	·····	·····	·····	·····	·····
18······	0.707	0.738	·····	0.9577	0.458	1.458	0.478	1.522	·····	·····	·····	·····	·····
19······	0.688	0.717	·····	0.9599	0.473	1.447	0.493	1.507	·····	·····	·····	·····	·····
20······	0.671	0.698	·····	0.9619	0.488	1.436	0.507	1.493	·····	·····	·····	·····	·····
21······	0.655	0.680	·····	0.9638	0.501	1.427	0.520	1.481	·····	·····	·····	·····	·····
22······	0.639	0.662	·····	0.9655	0.513	1.418	0.531	1.469	·····	·····	·····	·····	·····
23······	0.626	0.647	·····	0.9670	0.525	1.409	0.543	1.457	·····	·····	·····	·····	·····
24······	0.612	0.632	·····	0.9684	0.535	1.401	0.552	1.447	·····	·····	·····	·····	·····
25······	0.600	0.619	·····	0.9697	0.545	1.394	0.562	1.438	·····	·····	·····	·····	·····

In order to control the manufacturing process continuously, take samples by the method described above and record the results in control chart. In this case, the aim is to control the manufacturing process; therefore, if a dot falls outside the control limits, you need to identify the cause and make a suitable modification.

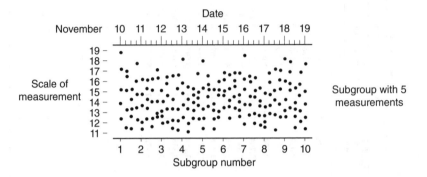

The above chart is used in the first stage of the explanation of control charts. A set of 185 measurements taken from the past record is divided into 37 groups with five measurements each. It is just an example for explanation.

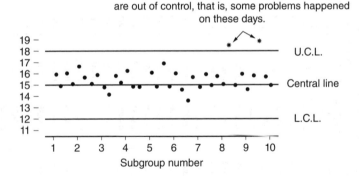

To make a control chart you have to calculate the values of the central line and the control limits. The value of the central line is obtained as the mean of the means of subgroups, and the control limits are obtained by calculating the standard deviation. Each point shows the mean of each subgroup. A special symbol is used for points falling outside the limits. In this example the mean of five points on a vertical line is plotted, but other statistics can be used in a similar manner.

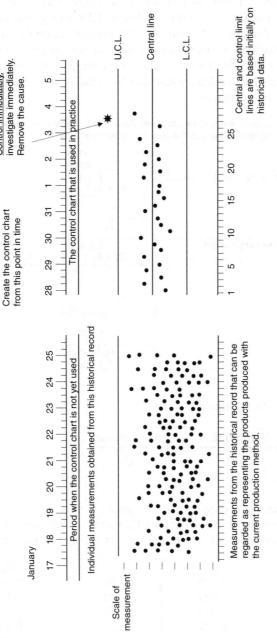

Control chart for controlling the quality of products during the production process. (This is a control chart for the means.) Each point represents the mean of each sample of size *n*.)

The explanation above is all about control charts for the mean and the standard deviation; however, you can also make other control charts such as for the range, the fraction defective, and the number of defectives. In these cases, the calculation method is almost the same as in the previous example, but different statistics and coefficients are used to determine the control limits. In the table of coefficients, the ones for the calculation of the range control chart are also included. The formulae to calculate statistics from the available measurements are shown in the following table.

Formulae for the case in which you are not given the standard of quality and you wish to analyze the past manufacturing record to create an initial control chart.

	Control line	Control limits
Mean control chart using sample standard deviations	$\bar{\bar{X}}$	$\bar{\bar{X}} \pm A_1\bar{\sigma}$ $\bar{\sigma} = \dfrac{\sigma_1 + \sigma_2 + \sigma_3 + \cdots + \sigma_m}{m}$ $A_1 = \dfrac{3}{c_2\sqrt{n}}$
Mean control chart using sample ranges	$\bar{\bar{X}}$	$\bar{\bar{X}} \pm A_1\bar{R}$ $\bar{R} = \dfrac{R_1 + R_2 + R_3 + \cdots + R_m}{m}$ $A_2 = \dfrac{3}{c_2\sqrt{n}}$
Sigma control chart	$\bar{\sigma}$	$B_3\bar{\sigma}$ and $B_4\bar{\sigma}$ $\bar{\sigma} \pm \dfrac{3\bar{\sigma}}{c_2\sqrt{n}}$
Range control chart	$\bar{R}$	$D_3\bar{R}$ and $D_4\bar{R}$ $\bar{R} \pm 3\sigma_R$

$$\sigma_{\bar{X}} = \frac{\sigma'}{\sqrt{n}}; \sigma' = \bar{R} \times \text{multiplier}$$

$$\bar{\bar{X}} \pm 3\sigma_{\bar{X}} = \bar{\bar{X}} \pm \frac{3\bar{R}\,\text{multiplier}}{\sqrt{n}}; K_n = \frac{3\bar{R}\,\text{multiplier}}{\sqrt{n}}$$

$$\therefore \bar{\bar{X}} \pm K_n\bar{R}$$

Table of coefficients or multipliers to determine the control limits for the mean using the average value of sample ranges ($\bar{R}$), where the coefficients K_n are the same as the coefficients A_2 in the table of coefficients on page 85.

n	K_n	n	K_n
2	1.880	9	0.337
3	1.023	10	0.308
4	0.729	11	0.285
5	0.577	12	0.266
6	0.483	13	0.249
7	0.419	14	0.235
8	0.373	15	0.223

Formulae to create control charts for controlling quality in a production process when standard values $\bar{X}$, σ', and R_n' $(R_n'=d_2\sigma')$ are known

	Central line	Control limits
Control chart for means	$\bar{X}'$	$\bar{X} \pm A\sigma'$, $\bar{X} \pm \frac{3\sigma'}{\sqrt{n}}$
Control chart for standard deviations	$C_2\sigma'$	$B_1\sigma'$ and $B_2\sigma'$ $C_2\sigma' \pm \dfrac{3\sigma'}{\sqrt{2n}}$
(When n is larger than 25, C_2 can be set to 1.0.)		
Control chart for ranges (when n is larger than 15)	$d_2\sigma'$	$D_1\sigma'$ and $D_2\sigma'$ $d_2\sigma' \pm 3\sigma_R$

There are two main uses of a control chart. One is to examine whether there exists or has existed a state of control in the current manufacturing process. The other is to maintain the state of control. The most important purpose of a control chart is to establish an operational standard for quality control of the manufacturing process and so establish the basis for systematic actions to modify the process. This aim is achieved by the combination of the above two uses.

From the historical record of the manufacturing process accumulated in factories, it is possible to ascertain the control limits and the condition of quality for those periods and also the variation in quality for the manufacturing method used at that time. We can often learn from past experience what we need in the future.

In order to estimate the stability of a manufacturing process, at least 25 successive dots are required, showing the states of the subgroups within the control limits. Otherwise, it is risky to make a judgment that the process is under the control. On the other hand, when one dot falls outside the control limits, then you can conclude that it is not under the control.

Knowledge about the controlled state is quite important. For example, it is useful when a manager seeks to establish a quality control system in a company for the first time. In order to determine what you need to make improvements, you need to know the original state of the factory and to know what happened after the introduction of the quality control system. The material purchasing department is also required to know about the state of control. They need to know whether purchased goods have stable quality. The stability of products is strongly affected by the quality of the materials used. For this reason, people in the material purchasing section will demand information about the quality of the materials from the vendor in advance or will do sampling inspection of the incoming materials themselves.

The other use of a control chart is to maintain the controlled state without stopping production. If it is possible to record product quality continuously, this will help us detect causes immediately after a quality change has occurred. In the process of making a control chart, the initial quality standard to be used in future operations is established by the design engineers or by the quality control engineers. For the latter case, the quality standard is found from the records of past production and inspection. In this way the control limits are specified for each production process and each quality characteristic. The control limits are set above and below the central line at distances determined by the sample size. The percentage of products expected to be within the limits should be taken into account in determining the control limits. At the same time, the expected percentage of defects can also be estimated.

Of course, before doing this, we have to make sure that the information and instructions are fully understood by everyone involved and are actually being applied in their work. Also, for both inspections in the factory and inspections for quality control, the inspection method, inspection instruments, instrumental devices, and inspectors should all be checked to ensure they are fit for purpose.

Now let us proceed to the stage of applying sampling inspection plans, which are determined on the basis of the research results by the quality control engineers. When applying the plan, we should take samples regularly at the same time interval, for example, once an hour, once per shift, or once per day, to compare the conditions of the manufacturing process on an ongoing basis. For each sample of measurements obtained by inspection, calculate the mean and the range or the mean and the standard deviation or the percentage of defects and plot them on the control chart. If there is one dot which falls outside the control limits, it shows that there is a cause that is producing a change in quality during the manufacturing process. We need to identify and eliminate this cause.

Regular reports based on the control chart should be done not only for the manager but also for production monitors responsible for the maintenance of product quality.

If there is a tendency to obtain points plotting outside the control limits, then the department of quality control reports to the department of manufacturing, upon which the department of manufacturing will take immediate action to rectify the situation. In this case the people from the department of quality control often work with the supervisor of the department of manufacturing to try to find the cause and to clarify the characteristics of the cause, and, if there is a need to change the process or an instrument, then to decide the level of changes.

Normally, if the department of quality control finds data points outside the limits, they immediately warn the manager of the manufacturing department and the direct operational supervisor. If possible, they send a copy of the warning to the department of production engineering and also to the department of design engineering, because there is a possibility that the problem is related to the department of design engineering. It is the responsibility of the engineers in the manufacturing department to solve the problem raised by the quality control department. In order to establish an effective organizational control system, the highest priority for the manufacturing department should be to solve the problem as soon as possible. The situation of a manufacturing department faced with this problem is similar to that of a driver whose car jumps off the pavement. He must take immediate preventive action to avoid danger.

However, it doesn't mean he needs to take action without a plan. He needs to:

1. clearly understand the problems highlighted by the warning,
2. establish the root causes,
3. analyze them,
4. plan the best solution,
5. carry out the plan, and
6. examine the results carefully to make sure that the problem will not occur again.

The supervisor of the process should report to his immediate superior what has been done to ensure the resolution of problems, and the results.

If a deviation from the control limits has occurred, the sampling method should be modified in order to check the quality of the succeeding lots more carefully and confirm a satisfactory quality level. If the quality is still unacceptable, then the general manager should be advised to stop the operational process until the situation improves. If the degradation in quality is serious, then operation of all processes should be stopped until it is clearly demonstrated that the cause of the problems has been rectified. Even after the operation has re-started, samples should be taken frequently to ensure that production has truly returned to a controlled state. In order to ensure this, 25 consecutive samples are required. If the process has returned to control, then sampling can revert to the original or an easier method.

Since the control limits provide the basis for action, they are sometimes called "action limits".

By repeating the above procedure of identifying and removing assignable causes for the variation or the change in quality and ensuring that the same problem never recurs, the manufacturing process approaches the ideal state of quality control.

Of course, even if quality is in control, it is not always satisfactory. That is the case when all the dots are within the control limits but lying close to one limit line. In order to modify this situation, one of the following two actions should be taken.

1. Shift the control standard by making a fundamental change to the production process.
2. Adjust the center of the specification limits to the center of the current state.

The effectiveness and usefulness of quality control is that it is possible to take action on the basis of the information gained by the activity of seeking the truth.

The most effective way of employing quality control is to apply the control chart to the parts of the manufacturing process that may have the most serious problems. It helps you to make timely modifications to this part of the process, and as a result overall efficiency will continuously improve. As production progresses day by day or month by month, the records of inspection and manufacturing and experience with quality control accumulate, enabling us to re-calculate the control limits and to find new values that lead to a more desirable and economical control set-up. The control chart method aims to continuously improve the conditions of manufacturing by the detection and elimination of assignable causes that disturb the conditions, and the goal of control is to eliminate all assignable causes using every means possible. So, the ongoing use of control charts and other appropriate methods make the whole manufacturing process as efficient and effective as possible.

What I have described above relates to explaining a control chart and some important things about its creation and use. That's because this aspect of quality control is directly related to the general management of companies. However, the function of the department of quality control is not limited to creating and maintaining control charts and carrying out sampling inspections. It has another important role: quality investigation.

The purpose of a quality investigation is to make a thorough investigation of the knowledge related to the products in question, including technical knowledge of the products and knowledge of their manufacturing process, that is, to investigate the current status of production by re-analyzing the methods or operations and studying the accuracy of machines and instruments. Usually, the products chosen for the investigation are selected from the company's ongoing mass production. The number of investigations depends on the current status of control, and on the availability of people to investigate. It should be done at least once a year, but generally it is desirable to do so more frequently.

The investigation is carried out by a committee comprising members of the departments of quality control, inspection, engineering, and manufacturing (production engineers are also included among committee members). Members are appointed on the recommendation of the managers of each department. Usually, this selection is done for every investigation.

The department of quality control is responsible for submitting to the director the title of the proposed investigation and the proposed beginning and end dates. Generally, work commences after the director approves this proposal.

It is necessary to allow sufficient time for members of the committee to prepare documents with information related to the product from their own standpoints, and to provide the information to other members in advance. For example, the department of engineering may think it better to change the technical standard, technical specification, or the specification for the materials, if they have some basic knowledge of the product. Also every member should give some thought to the items to be discussed by all committee members.

The department of quality control should check the manufacturing conditions of the factory to ensure that all operations are instructed in detail and that the conditions are consistent with the new standard for manufacturing techniques. Also, it should check whether there are effective operating standards that are determined by the department of manufacturing but of which the department of engineering is unaware. The department of quality control should analyze customer complaints, the results of inspecting finished product from the factory, and the results from current sampling inspection.

People in the department of manufacturing should report the status of machines, instruments, materials, stocks, and difficulties encountered during manufacturing, and the possible status of the quality of the products.

The department of inspection should monitor and make recommendations about the status of instruments, the accuracy of instruments, and the number of modifications required and any issues arising during the inspection.

On the scheduled day, the committee meets and the preliminary reports are analyzed to see how to address the issues that arose during the investigation. Apart from committee members, any person who has specialized knowledge of a particular area should be invited. The meeting is not just held in a room: members also go to the factories and study each stage of the operation and the current status of inspection in the factories and in the department of quality control.

At the conclusion of the investigation, the committee writes a progress report that is signed by the committee members, indicating that each member has equal responsibility for the investigation.

Usually, the investigation report consists of three parts. The first part goes to the CEO. It is written in summary form, describing the competence of the department of manufacturing and the department of engineering, and the fitness for purpose of the products. The committee might uncover problems requiring changes to be made. In this case, the places where the problems are occurring and suggestions for possible policies for solving the problems should be itemized. The department responsible for solving the problems should also be documented.

The second part of the report is sent to the department of engineering. It consists of a detailed analysis of the problems that the committee judges require action by the department of engineering. The problems written in part one should be elaborated in greater detail with accompanying data and suggestions for action to solve the problems should be given.

The third part of the investigation report is sent to the department of manufacturing, and problems to be solved in the department should be clearly spelt out. As in the report sent to the department of engineering, suggestions should be given about how to solve the problems.

This committee is the actual investigating body. Therefore, it is not authorized to order any department to take action. However, having read the committee's report, a conscientious manager might sometimes ask for a report on the outcomes of the suggested actions in each department. The CEO may request regular status reports about final results, progress, or unsolved problems. Moreover, he may sometimes call investigation team members and order them to evaluate the outcomes by showing them the information from the status reports.

The third important function of the department of quality control is to investigate products or items in response to complaints about quality. This is as important as the other two controls: statistical quality control and quality investigation. It doesn't matter where the complaint came from, dissatisfaction with the quality of items indicates that there are undetected avoidable defects during production. For this reason, it is worth analyzing the problems and clarifying the cause of the complaint. If the investigation is carried out extensively in the factory, in the department of design, and in other departments, considerable improvement of quality will be achieved. Sometimes, investigating a customer complaint can be a shortcut to achieving better quality control. However, investigating such complaints shouldn't replace the planned program of quality control to improve and maintain the controlled state. Such an investigation is useful as a review of whole production system.

Appendix: Statistical Tables for Sampling Inspection

SL Table-1.0, DL Table 5, SA Table-10.0, DA Table-1.0
(From "Sampling Inspection Tables" by Dodge, H.F. and Romig, H.G.)

SL Table – 1

Lot tolerance percent defects in a lot = 1.0%

Process mean %	0–.010			0.011–.10			0.11–.20			0.21–.30			0.31–.40			0.41–.50		
Lot size	n	c	AOQL (%)	n	c	AOQL (%)	n	c	AOQL (%)	n	c	AOQL (%)	n	c	AOQL (%)	n	c	AOQL (%)
1 – 120	All	0	0	All	0	0	All	0	0	All	0	0	All	0	0	All	0	0
121 – 150	120	0	.06	120	0	.06	120	0	.0	120	0	.06	120	0	.06	120	0	.05
151 – 200	140	0	.08	140	0	.08	140	0	.0	140	0	.08	140	0	.08	140	0	.08
201 – 300	165	0	.10	165	0	.10	165	0	.1	165	0	.10	165	0	.10	165	0	.10
301 – 400	175	0	.12	175	0	.12	175	0	.1	175	0	.12	175	0	.12	175	0	.12
401 – 500	180	0	.13	180	0	.13	180	0	.1	180	0	.13	180	0	.13	180	0	.13
501 – 600	190	0	.13	190	0	.13	190	0	.1	190	0	.13	190	0	.13	303	1	.14
601 – 800	200	0	.14	200	0	.14	200	0	.1	330	1	.15	330	1	.15	330	1	.15
801 – 1000	205	0	.14	205	0	.14	205	0	.1	335	1	.17	335	1	.17	335	1	.17
1001 – 2000	220	0	.15	220	0	.15	360	1	.1	490	2	.21	490	2	.21	610	3	.22
2001 – 3000	220	0	.15	375	1	.20	505	2	.2	630	3	.24	745	4	.26	870	5	.26
3001 – 4000	225	0	.15	380	1	.20	510	2	.2	645	3	.25	880	5	.28	100	6	.29
4001 – 5000	225	0	.16	380	1	.20	520	2	.2	770	4	.28	895	5	.29	112	7	.31
5001 – 7000	230	0	.16	385	1	.21	655	3	.2	780	4	.29	102	6	.32	126	8	.34
7001 – 10,000	230	0	.16	520	2	.25	660	3	.2	910	3	.32	115	7	.34	150	1	.37
10,001 – 20,000	390	1	.21	525	2	.26	785	4	.3	104	6	.35	140	9	.39	198	1	.43
20,001 – 50,000	390	1	.21	530	2	.26	920	5	.3	130	5	.39	159	1	.44	257	1	.48
50,001 – 100,000	390	1	.21	167	3	.29	1040	6	.3	142	9	.41	212	1	.47	315	2	.50

n = the size of a sample, 'All' means all items in the lot

c = Acceptable number of defects allowed in the sample

AOQL = average outgoing quality limit

DL Table – 5

Lot tolerance percent defects = 5.0

Lot size	0–.05 1st n₁	1st c₁	2nd n₂	2nd m₁+n₂	AOQL %	.06–.50 1st n₁	1st c₁	2nd n₂	2nd m₁+n₂	c₂	AOQL %	.51–1.0 1st n₁	1st c₁	2nd n₂	2nd m₁+n₂	c₂	AOQL %	1.01–1.55 1st n₁	1st c₁	2nd n₂	2nd m₁+n₂	c₂	AOQL %	1.51–2.00 1st n₁	1st c₁	2nd n₂	2nd m₁+n₂	c₂	AOQL %	2.01–2.50 1st n₁	1st c₁	2nd n₂	2nd m₁+n₂	c₂	AOQL %
1–30	All 30	0	—	—	.49	All 30	0	—	—	—	.49	All 30	0	—	—	—	.49	All 30	0	—	—	—	.49	All 30	0	—	—	—	.49	All 30	0	—	—	—	.49
31–50	38	0	—	—	.59	38	0	—	—	—	.59	38	0	—	—	—	.59	38	0	—	—	—	.59	38	0	—	—	—	.59	38	0	—	—	—	.59
51–75	44	0	21	65	.64	44	0	21	65	1	.64	44	0	21	65	1	.64	44	0	21	65	1	.64	44	0	21	65	1	.64	44	0	21	65	1	.64
76–100	49	0	25	75	.84	49	0	26	75	1	.84	49	0	26	75	1	.84	49	0	51	100	2	.84	49	0	51	100	2	.91	49	0	51	100	2	.91
101–200	50	0	30	80	.91	50	0	30	80	1	.91	50	0	55	105	2	.0	50	0	55	105	2	.0	50	0	80	130	3	.1	50	0	100	150	4	.1
201–300	55	0	30	85	.92	55	0	55	110	2	.1	55	0	80	135	3	.1	55	0	80	135	3	.1	55	0	100	155	4	.2	85	1	105	190	6	.3
301–400	55	0	30	85	.93	55	0	55	110	2	.2	55	0	90	135	3	.2	55	0	105	160	4	.2	85	1	120	205	6	.4	85	1	140	225	7	.4
401–500	55	0	30	85	.94	55	0	55	110	2	.2	55	0	115	170	4	.2	55	0	105	160	4	.2	85	1	120	205	6	.4	85	1	140	225	7	.4
501–600	55	0	30	85	.94	55	0	60	115	3	.2	55	0	85	140	4	.2	55	1	110	165	4	.2	85	1	145	230	7	.4	85	2	165	250	8	.5
601–800	55	0	35	90	.95	55	0	65	120	3	.3	55	0	85	140	4	.3	55	0	125	215	6	.3	90	1	170	250	8	.5	120	2	185	305	10	.6
801–1000	55	0	35	90	.95	55	0	65	120	3	.3	55	0	115	170	4	.4	90	1	150	240	7	.4	90	1	200	290	9	.6	120	2	210	330	11	.7
1001–2000	55	0	35	90	.98	55	0	95	150	4	.3	55	0	120	175	4	.4	90	1	185	275	8	.4	120	2	225	345	11	.9	175	4	260	435	15	.0
2001–3000	55	0	65	120	.2	55	0	95	150	4	.3	55	0	150	175	5	.5	120	2	190	300	9	.5	150	3	270	420	14	.1	205	5	375	580	21	.3
3001–4000	55	0	65	120	.2	55	0	95	150	4	.3	90	1	150	230	6	.6	120	2	210	330	10	.6	150	3	295	445	15	.3	230	6	420	650	24	.4
4001–5000	55	0	65	120	.2	55	0	95	150	4	.3	90	1	165	255	7	.8	120	2	255	375	12	.1	150	3	345	495	17	.3	255	7	445	700	25	.5
5001–7000	55	0	65	120	.2	55	0	95	150	4	.3	90	1	165	255	7	.8	120	2	260	380	12	.1	150	3	370	520	18	.3	255	7	495	750	28	.6
7001–10,000	55	0	65	120	.2	55	0	120	175	5	.5	90	1	190	280	8	.9	120	2	285	405	13	.1	175	4	370	545	19	.4	280	8	540	820	31	.7
10,001–20,000	55	0	65	120	.2	55	0	190	190	7	.5	90	1	190	280	8	.9	120	2	310	430	14	.2	175	4	420	595	21	.4	280	8	660	940	36	.8
20,000–50,000	55	0	65	120	.2	55	0	215	305	9	.7	90	1	215	305	9	.0	120	2	335	455	15	.2	105	5	485	690	25	.5	305	9	745	1050	41	.9
50,001–100,000	55	0	65	120	.2	90	1	240	330	10	.2	90	1	240	330	10	.1	120	2	360	480	16	.3	105	5	555	760	28	.6	330	10	810	1140	45	.0

n₁ = Size of the first sample; n₂ = Size of the second sample.

c₁ = Acceptable number of defects allowed in the first sample; c₂ = Acceptable total number of defects allowed in the first and second samples

AOQL = Average outgoing quality limit

SA Table – 10
Average outgoing quality limit = 10.0%

Process mean %	0 – 0.010			0.011 – 10			0.11 – 20			0.21 – 30			0.31 – 40			0.41 – 50		
Lot size	n	c	AOQL (%)	n	c	AOQL (%)	n	c	AOQL (%)	n	c	AOQL (%)	n	c	AOQL (%)	n	c	AOQL (%)
1 – 120	All	0	0	All	0	0	All	0	0	All	0	0	All	0	0	All	0	0
121 – 150	120	0	.06	120	0	.06	120	0	.0	120	0	.06	120	0	.06	120	0	.05
151 – 200	140	0	.08	140	0	.08	140	0	.0	140	0	.08	140	0	.08	140	0	.08
201 – 300	165	0	.10	165	0	.10	165	0	.1	165	0	.10	165	0	.10	165	0	.10
301 – 400	175	0	.12	175	0	.12	175	0	.1	175	0	.12	175	0	.12	175	0	.12
401 – 500	180	0	.13	180	0	.13	180	0	.1	180	0	.13	180	0	.13	180	0	.13
501 – 600	190	0	.13	190	0	.13	190	0	.1	190	0	.13	190	0	.13	303	1	.14
601 – 800	200	0	.14	200	0	.14	200	0	.1	330	1	.15	330	1	.15	330	1	.15
801 – 1000	205	0	.14	205	0	.14	205	0	.1	335	1	.17	335	1	.17	335	1	.17
1001 – 2000	220	0	.15	220	0	.15	360	1	.1	490	2	.21	490	2	.21	610	3	.22
2001 – 3000	220	0	.15	375	1	.20	505	2	.2	630	3	.24	745	4	.26	870	5	.26
3001 – 4000	225	0	.15	380	1	.20	510	2	.2	645	3	.25	880	5	.28	1000	6	.29
4001 – 5000	225	0	.16	380	1	.20	520	2	.2	770	4	.28	895	5	.29	1120	7	.31
5001 – 7000	230	0	.16	385	1	.21	655	3	.2	780	4	.29	1020	6	.32	1260	8	.34
7001 – 10,000	230	0	.16	520	2	.25	660	3	.2	910	3	.32	1150	7	.34	1500	10	.37
10,001 – 20,0000	390	1	.21	525	2	.26	785	4	.3	1040	6	.35	1400	9	.39	1980	14	.43
20,001 – 50,000	390	1	.21	530	2	.26	920	5	.3	1300	5	.39	1590	13	.44	2570	19	.48
50,001 – 100,000	390	1	.21	1670	3	.29	1040	6	.3	1420	9	.41	2120	15	.47	3150	23	.50

n = Sample size; "All" means all items in the lot

c = Acceptable number of defects allowed in the sample

p_t = Lot tolerance percent defectives corresponding to Consumers' risk $(P_c) = 0.10$

DA Table – 1

Average outgoing quality limit = 1.0 %

Lot size	0–.02						.03–.20						.21–.40						.41–.60						.61–.80						.81–1.00					
	n_1	c_1	n_2	n_1+n_2	c_2	p_t %	n_1	c_1	n_2	n_1+n_2	c_2	p_t %	n_1 (Trial 1)	c_1	n_2	n_1+n_2	c_2	p_t %	n_1	c_1	n_2	n_1+n_2	c_2	p_t %	n_1	c_1	n_2	n_1+n_2	c_2	p_t %	n_1	c_1	n_2	n_1+n_2	c_2	p_t %
1–25	All	0	–	–	–	–	All	0	–	–	–	–	All	0	–	–	–	–	All	0	–	–	–	–	All	0	–	–	–	–	All	0	–	–	–	–
26–50	22	0	–	–	–	7.7	22	0	–	–	–	7.7	22	0	–	–	–	7.7	22	0	–	–	–	7.7	22	0	–	–	–	7.7	22	0	–	–	–	7.7
51–100	33	0	17	50	–	6.9	33	0	17	50	–	6.9	33	0	17	50	–	6.9	33	0	17	50	1	6.9	33	0	17	50	1	6.9	33	0	17	50	1	6.9
101–200	43	0	22	65	1	5.8	43	0	22	65	1	5.8	43	0	22	65	1	5.8	43	0	22	65	1	5.8	43	0	22	65	1	5.8	47	0	40	90	1	5.4
201–300	47	0	28	75	1	5.5	47	0	28	75	1	5.5	47	0	28	75	1	5.5	55	0	50	105	2	4.9	55	0	50	105	2	4.9	55	0	50	105	2	4.9
301–400	49	0	31	80	1	5.4	49	0	31	80	1	5.4	55	0	60	115	2	4.8	55	0	60	115	2	4.8	55	0	60	115	2	4.8	60	0	80	140	3	4.5
401–500	50	0	30	80	1	5.4	50	0	30	80	1	5.4	55	0	65	120	2	4.7	55	0	65	120	2	4.7	60	0	95	155	3	4.3	60	0	95	155	3	4.3
501–600	50	0	30	80	1	5.4	50	0	30	80	1	5.4	60	0	65	125	2	4.6	60	0	65	125	2	4.6	65	0	100	165	3	4.2	65	0	100	165	3	4.2
601–800	50	0	35	85	1	5.3	60	0	70	130	2	5.3	60	0	70	130	2	4.5	65	0	105	170	3	4.1	65	0	105	170	3	4.1	70	0	140	210	4	3.9
801–1000	55	0	30	85	1	5.2	60	0	75	135	2	5.2	60	0	75	135	2	4.4	65	0	110	175	3	4.0	70	0	150	220	4	3.8	125	1	180	305	6	3.5
1001–2000	55	0	35	90	1	5.1	65	0	75	140	2	4.3	75	0	120	195	3	3.8	80	0	165	245	4	3.7	135	1	200	335	6	3.3	140	1	245	385	7	3.2
2001–3000	65	0	80	145	2	4.2	65	0	80	145	2	4.2	75	0	125	200	3	3.7	85	0	165	250	4	3.6	150	1	265	415	7	3.0	215	2	355	570	10	2.8
3001–4000	70	0	80	150	2	4.1	70	0	80	150	2	4.1	80	0	175	255	4	3.5	85	0	220	305	5	3.3	160	1	330	490	8	2.8	225	2	455	680	12	2.7
4001–5000	70	0	80	150	2	4.1	70	0	80	150	2	4.1	80	0	180	260	4	3.4	145	1	225	370	6	3.1	225	2	375	600	10	2.7	240	2	595	835	16	2.5
5001–7000	70	0	80	150	2	4.1	75	0	125	200	3	3.7	80	0	180	260	4	3.4	155	1	285	440	7	2.9	235	2	440	675	11	2.6	310	3	665	975	26	2.4
7001–10,000	70	0	80	150	2	4.1	80	0	125	205	3	3.6	85	0	180	265	4	3.3	165	1	355	520	8	2.7	250	2	585	835	13	2.4	385	4	785	1170	19	2.3
10,001–20,000	70	0	80	150	2	4.1	80	0	130	210	3	3.6	90	0	230	320	5	3.2	175	1	415	590	9	2.6	325	3	655	980	15	2.3	520	6	980	1500	24	2.2
20,000–50,000	75	0	80	155	2	4.1	80	0	135	215	3	3.6	95	0	300	395	6	2.9	250	2	490	740	11	2.4	340	3	910	1250	19	2.2	610	7	1410	2020	32	2.1
50,001–100,000	75	0	80	155	2	4.1	85	0	180	265	4	3.3	170	1	380	550	8	2.6	275	2	700	975	14	2.2	420	4	1050	1470	22	2.1	770	9	1850	2620	41	2.0

n_1 = Size of the first sample; n_2 = Size of the second sample

c_1 = Acceptable number allowed in the first sample; c_2 = Acceptable total number of defects allowed in the first and second samples

p_t = Lot tolerance percent defectives corresponding to Consumers' risk $(P_c) = 0.10$

Published

- HOEL… Introduction to Mathematical Statistics, 300 pages, ¥400
- Statistical Methods for Lifetime Testing and Estimation, 120 pages, ¥350
- Dodge and Romig…Tables for Sampling Inspection, 95 pages, ¥350
- QMC Sampling Inspection Handbook, 260 pages, ¥500

To be published

- Sequential Analysis in Statistics, by Columbia University statistical investigation team, ¥1800
- Statistical Figures and Tables, ¥800
- Study of Movement using Statistical Inference, ¥500
- Jessen… Survey Design, ¥500

Thoughts on *The Road to Quality Control – The Industrial Application of Statistical Quality Control* by Homer M. Sarasohn

W. H. Woodall

Virginia Tech, Blacksburg, USA

Introduction

When I was asked to write a short article on *The Road to Quality Control*, I had very limited knowledge of Homer Sarasohn. I had heard his name, knew he had been an electrical engineer from the US who had worked in Japan immediately after the end of WWII, and that was about it. I have since learned that his work in Japan to lead the restoration of radio and telephone communication was requested by General Douglas MacArthur. Sarasohn was only 29 years old at the time with an undergraduate degree in physics.

After reading this translation of Sarasohn (1952) and learning more about his work in Japan, it is my view that Sarasohn's contributions to Japanese post-war industrial success through the Industry Branch of the occupation army's Civil Communications Section (CCS) have been greatly underappreciated. His work in Japan in 1946–1950 in coordinating, from rubble, the building of what became Japan's most successful electronics companies preceded the first quality-related visit by W. E. Deming in 1950 and the first visit by Joseph Juran in 1954. In order to be successful, Sarasohn learned and then taught classes in Japanese. His *The Fundamentals of Industrial Management* (Sarasohn & Protzman, 1949) with Charles Protzman, a senior production manager with AT&T, and *The Road to Quality Control* were taught to managers from a wide range of companies and published in Japanese in 1949 and 1951, respectively.

The Road to Quality Control is written in a very readable style and not so detailed as to cause readers to lose sight of the important issues. The writing is excellent and sometimes elegant. For example, on p. 91, Sarasohn writes,

The Road to Quality Control: The Industrial Application of Statistical Quality Control by Homer M. Sarasohn,
First Edition. Translated by N.I. Fisher & Y. Tanaka from the original Japanese text published
by Kagaku Shinko Sha with a historical perspective by W.H. Woodall and a historical context by N.I. Fisher.
© 2019 John Wiley & Sons Ltd. Published 2019 by John Wiley & Sons Ltd.

The effectiveness and usefulness of quality control is that it is possible to take action on the basis of the information gained by the activity of seeking the truth.

Quality Management

It is important to note that Sarasohn presents the technical aspects of sampling methods and statistical process control (SPC) within the context of an overall quality management system. It came as a great surprise to me to find his quality management principles closely match many of those of total quality management (TQM) which did not become widely accepted in the US until the quality revolution of the 1980s. Thus, from a quality management perspective, the book was in many ways decades ahead of its time.

In an interview with Sarasohn, Myron Tribus (1988) characterized Sarasohn's management system as representing the best philosophy of management. Sarasohn said he based his management system upon his own experience, but since he was starting from scratch with free rein in post-war Japan, he incorporated his view of how management *ought* to work.

With respect to the importance of quality, Sarasohn states on p. 3 a somewhat radical idea at the time,

> There can be no doubt that quality is the most essential foundation for a company to be successful.

To illustrate how closely Sarasohn's principles match those of TQM, the primary characteristics of TQM are given below followed by some relevant quotes from his book.

1. **A focus on customer satisfaction.**

> The most important management principle for maintaining a high level of product is to establish company-wide determination to serve their customers. (p. 1)

2. **Management leadership for quality.**

> Managers should take the initiative in related activities. They should show with their behavior that quality is their company motto, and they also should try to ensure that the workplace environment is conducive to maintaining the highest possible quality. (p. 1)

3. **Quality is the responsibility of all employees.**

> It is planned so that no-one in the company is indifferent to maintaining quality. (p. 76)

> ...treatment and training of employees is very important. (p. 1)

4. **Company focuses on continuous improvement.**

He wrote,

> ...we not only can, but we must minimize, although not exactly, the amount of variability in the quality of production in a unit or lot ... This is the purpose of quality control. (p. 17)

Sarasohn acknowledged the importance of preventing out-of-specification product, and goes on to write,

> ...quality control is used to reduce the variation in product quality. (p. 76)

A key driver of continuous improvement is the realization that meeting specifications is not sufficient. Instead one must reduce variation around a target value. This was a key message of Deming and, later, Genichi Taguchi. One of the weaknesses of TQM, however, that led to Six Sigma's emergence was that profit and cost/benefit tradeoffs were not adequately considered. Sarasohn understood costs and discusses, on p. 4, that quality standards should be set so that the product remains commercially viable.

In addition, Sarasohn also makes some points which are now regarded as characterizing the fundamental principles of statistical thinking (ASQ Statistics Division, 1996).

1. **All work occurs in a system of interconnected processes.**

> From the moment when the materials arrive, to the shipping of products, there's no error of quality that can be permitted in any stage of the process. (p. 75)

2. **Variation exists in all processes.**

> From a philosophical point of view, it can be postulated that no two things in the world are the same. All events and things can be similar but there must be *some* differences. (p. 19; his emphasis)

Therefore, even in the same manufacturing organization a series of products cannot be exactly the same. (p. 8)

3. **Understanding and reducing variation are the keys to success.**

Again, as listed above under TQM, Sarasohn wrote,

...quality control is used to reduce the variation in product quality. (p. 76)

The only quality management idea that doesn't reflect current quality management thought is to set up a quality inspection division which is to operate independently from manufacturing. There may have been valid reasons for this approach in post-war Japan, but it can lead to conflict. As Sarasohn states on p. 5, it makes the duty of inspection "similar to that of prosecution".

Use of Acceptance Sampling

Sarasohn emphasizes the use of acceptance sampling, giving its advantages over 100% inspection on p. 63. Bypassing inspection entirely in post-war Japanese manufacturing was not realistic, owing to the poor yields. Sarasohn advises there be no inspection only when the proportion of non-conforming products is quite low and one is confident that the process will remain stable.

Deming (1952) devoted his sixth and last lecture in Japan to acceptance sampling. He explained single and double sampling methods in some detail, but criticized acceptance sampling as an approach that raises costs and does not significantly improve quality. He stated that a purpose of acceptance sampling should be to force the use of control charts. Acceptance sampling fell out of favor during the 1980s. This was in part due to criticism by Deming (1986, p. 133), who wrote, for example,

Incredibly, courses and books in statistical methods still devote time and pages to acceptance sampling.

Unnecessary inspection is a waste of time and money, but, as discussed by Vardeman (1986), there are many cases in practice where sampling inspection remains useful.

Control Chart Methods

The standard control chart calculations for the X-bar chart, the R-chart, the S-chart, and the p-chart are presented. There is a focus on simplifying the calculations, as one would expect for the time.

Sarasohn carefully distinguishes between what are now called Phase I and Phase II in process monitoring. In Phase I, the sole focus of Shewhart (1931, 1939), one analyzes historical data to learn about the process and to assess its stability and capability. Jones-Farmer et al. (2014) provide an overview of Phase I issues and methods. In Phase II, however, one uses control limits obtained from Phase I analysis to maintain and monitor process stability. This is an important distinction both practically and theoretically. On this topic, Sarasohn writes,

> There are two main uses of a control chart. One is to examine whether or not there exists or has existed a state of control in the current manufacturing process. The other is to maintain the state of control. (p. 89)

It is also clear that Sarasohn understood the key practical concept of rational subgrouping, i.e. the choice of sampling scheme that is most likely to result in the detection of any removable assignable cause of variation. On p. 79, he presents this idea in terms of solving a problem by defining lots to be sampled in terms of machine used rather than by day manufactured. Sarasohn offers a considerable amount of practical advice. For example, he recommends moving quality evaluation upstream in the production process. He warns practitioners to pay attention to measurement error. He also recommends taking smaller samples more frequently instead of larger samples less frequently.

Sarasohn knew of Shewhart's work and had access to Shewhart's 1931 book. Charles Protzman was a long-time Western Electric employee, so this might be expected. His statistical notation follows that of Shewhart (1931) to a large extent and may be confusing to readers since it is considerably different from what is used now. For example, both Sarasohn and Shewhart used the symbol σ_σ to represent the standard deviation of the sample standard deviation. The symbol σ_S reflects current practice.

My only technical criticism relative to process monitoring is that Sarasohn refers to tolerance limits as "control limits" on p. 83. Tolerance limits might match control limits closely, but only if one is monitoring the mean of a process and the sample size is one at each time period.

Theory vs. Practice

On p. 61 Sarasohn warns against putting an emphasis on the theoretical side of statistical quality control. He writes,

> There are some scientists who placed unreasonable emphasis just on the theoretical side of quality control. It is an undesirable attitude. If this attitude prevails, then there's no hope of applying quality control to get good results. ... Quality control is one kind of application of mathematics

but fundamentally, it's not an area for theoreticians to work in. Practical application is the point to be emphasized.

Based on Fisher (2009), this view was likely a reaction to some members of the Union of Japanese Scientists and Engineers (JUSE) who began to overemphasize the importance of theory. SPC is unlikely to be successful unless it is used within an effective management structure. A similar negative view regarding the usefulness of theoreticians is expressed by Wheeler (2017), but, as discussed in Woodall (2017), theory can complement practice.

Some Other Books of the Era

The earlier books on statistical quality control by Grant (1946) and Enrick (1948) go into far greater detail on statistical methods than does *The Road to Quality Control*. In the 1954 second edition of his book, N. L. Enrick devotes only five pages to management organization. In the 1964 third edition of his book, Eugene Grant does not discuss any management issues whatsoever. Unfortunately, statistical aspects of quality became separated from managerial aspects early on in the literature and largely remain separated. Rice (1947) seems to have taken a broader view by including some managerial aspects.

For some historical perspective, Deming's *Elementary Principles of the Statistical Control of Quality: A Series of Lectures*, which contains the material he taught in Japan, first appeared in 1950. He used the American War Standards Z1.1–Z1.3, published in 1941–1942, in his teaching (American Standards Association, 1941). The first edition of Juran's *Quality Control Handbook* was published in 1951 and the *Western Electric Handbook* in 1956.

Conclusions

It is impossible to assess accurately the relative impact of Sarasohn, Deming, and Juran on post-war Japanese manufacturing success. Deming (1952) focused almost entirely on statistical issues and did not offer a roadmap for organizing a quality management system. According to Kolesar (2008), Juran was invited to lecture in Japan on the management of quality, owing to a perceived overemphasis in Japan on the use of statistics.

Deming and Juran are given the lion's share of the credit for the emphasis on quality that contributed to Japan's post-war industrial success. Sarasohn's work has garnered relatively little attention, even though he contributed directly to Japanese industry through the CCS. In addition, his progressive quality management system, which included an appropriate focus on the use of statistics, was promulgated throughout Japanese industry prior to the arrival of Deming and Juran.

If Sarasohn's two books had been published in English in the 1950s, he might very well be held to the same high level of esteem as are Deming and Juran. His work in helping to create the highly successful Japanese electronics industry is well documented. His forward-thinking quality-management ideas and his understanding of industrial statistical practice point to an exceptional level of genius that should not be overlooked.

References

American Standards Association (1941). *American War Standards Z1.1–Z1.2: Guide for Quality Control and Control Chart Method of Analyzing Data.* New York: American Standards Association.

ASQ Statistics Division (1996). *Glossary & Tables for Statistical Quality Control,* 3e. Milwaukee, WI: ASQ Quality Press Publications.

Deming, W.E. (1952). *Elementary Principles of the Statistical Control of Quality: A Series of Lectures.* Tokyo: Nippon Kagaku Gijutsu Remmei.

Deming, W.E. (1986). *Out of the Crisis.* Cambridge, MA: Massachusetts Institute of Technology, Center for Advanced Engineering Study.

Enrick, N.L. (1948). *Quality Control.* New York: The Industrial Press.

Fisher, N.I. (2009). Homer Sarasohn and American involvement in the evolution of quality Management in Japan, 1945–1950. *International Statistical Review* 77 (2): 276–299.

Grant, E.L. (1946). *Statistical Quality Control.* New York: McGraw-Hill, Inc.

Jones-Farmer, L.A., Woodall, W.H., Steiner, S.H., and Champ, C.W. (2014). An overview of phase I analysis for process improvement and monitoring. *Journal of Quality Technology* 46 (3): 265–280.

Kolesar, P.J. (2008). Juran's lectures to Japanese executives in 1954: A perspective and some contemporary lessons. *Quality Management Journal* 15 (3): 7–16.

Rice, W.B. (1947). *Control Charts and Factory Management.* New York: Wiley.

Sarasohn, H.M. (1952). *The Industrial Application of Statistical Quality Control.* (Translated into Japanese by Gonta Tsunemasa.). Tokyo: Kagaku Shinko Sha. (The 2017 English version is available.).

Sarasohn, H.M. and Protzman, C.W. (1949). The Fundamentals of Industrial Management. Civil Communications Section Management Course (in Japanese). (1998 English translation available at valuemetrics.com.au/resources003.html, accessed 6/13/2017).

Shewhart, W.A. (1931). *Economic Control of Quality of Manufactured Product.* Van Nostrand. (Reprinted by American Society of Quality Control, 1980.).

Shewhart, W.A. (1939). *Statistical Method from the Viewpoint of Quality Control* (ed. W.E. Deming). Washington, D.C: Graduate School of the Department of Agriculture (Republished in 1986 by Dover Publications, Inc., Mineola, NY.).

Tribus, M. (1988). Interview with Homer Sarasohn. blog.deming.org/2017/02/myron-tribus-interview-of-homer-sarasohn/, accessed 6/13/2017.

Vardeman, S.B. (1986). The legitimate role of inspection in modern SQC. *The American Statistician* 40 (4): 325–328.

Wheeler, D.J. (2017). Statistical process control. *ASQ Statistics Division Newsletter* 36 (1): 21–28.

Woodall, W.H. (2017). Bridging the gap between theory and practice in basic statistical process monitoring. *Quality Engineering* 29 (1): 2–15.

Homer Sarasohn and American Involvement in the Evolution of Quality Management in Japan, 1945–1950*

N.I. Fisher[1, 2]

[1] University of Sydney, School of Mathematics & Statistics, Australia
[2] ValueMetrics Australia, Australia

Summary

The history of Quality Management, and of the role of Statistics in Quality Management, is inextricably bound to the reconstruction of Japan immediately following the Second World War, and then to developments in the United States over three decades later. Even though these periods are, in societal history, just moments ago, yet there is profound lack of agreement about what was actually done, and who should be recognized for their contributions. This paper draws on historical materials recently made publicly available in order to clarify what actually took place between 1946 and 1950, and in particular the contribution of a remarkable engineer, Homer Sarasohn.

Keywords *Charles Protzman; Civil communications section; General Douglas MacArthur; Homer Sarasohn; Statistical process control; Total Quality Control; Union of Japanese Scientists and Engineers; W. Edwards Deming.*

1 Introduction

The history of Quality Management is inextricably bound to the reconstruction of Japan immediately following the Second World War, and then to developments in the United States over three decades later. Even though these periods are, in societal history, just moments ago, yet there is profound lack of agreement about what was actually done, and who should be recognized for their contributions, despite the considerable efforts of two outstanding

*Originally printed in *International Statistical Review* (2009), 77, 2, 276–299 doi:10.1111/j.1751-5823.2008.00065.x

The Road to Quality Control: The Industrial Application of Statistical Quality Control by Homer M. Sarasohn, First Edition. Translated by N.I. Fisher & Y. Tanaka from the original Japanese text published by Kagaku Shinko Sha with a historical perspective by W.H. Woodall and a historical context by N.I. Fisher.

scholars. None of the original contributors is still alive, so we have very little information available in the form of primary resources (with one outstanding exception, mentioned below).

The purpose of this paper is two-fold:

a. To set out what appear to be largely undisputed data—in other words, what might reasonably be accepted as factual.
b. To provide an interpretation based on the factual information and other more subjective material from secondary sources, including the author's own experiences from interactions with primary contributors.

The discussion draws heavily on the work of the two leading scholars in the area: Myron Tribus, who has studied historical developments very broadly; and Kenneth Hopper, who has devoted a lifetime to studying what happened in post-war Japan, and subsequently in the USA. The perspective of this paper is that of a statistical scientist, so there is explicit focus on what is often termed "Statistical Quality Control".

Key sources of information for this paper, including articles and correspondence recently made available by the late Homer Sarasohn's daughter, Lisa Sarasohn, are provided in Section 6.

For more information about developments in Quality Management and Quality Practice before 1945 and after 1950, the reader is referred to Fisher & Nair (2009) and references therein.

2 Events Prior to 1945

2.1 Data

Tsutsui (1996) reported on pre-1945 work related to Quality Control, based on studying Japanese resources:

- 1931: A Japanese engineer, Ishida Yasushi, studied the latest American techniques and developed a distinctive "scroll" (*makimono*) system of control charts for use in Tokyo Shibaura Electric factories.
- 1934: Kiribuchi Kanzō published a monograph on the use of statistical analysis in the production process to ensure conformity to standards.
- 1942: Ishida Yasushi and Kitagawa Toshio (from Kyushu University) published a Japanese translation of E. S. Pearson's industrial statistics book (Pearson, 1935).
- 1943: A public–private "research group" (*kenkyū tonarigumi*) was formed under the auspices of the Technology Agency (Gijutsu-in) to study mathematical approaches to mass production.

(Quoted largely verbatim from Tsutsui (1996), where the specific references in Japanese can be found. Important note: Japanese-language references cited by Tsutsui *have not been checked directly* by the author.)

3 1945–1947

3.1 Data

3.1.1 Civil Communications Section (CCS) activities

At the end of 1945, General Douglas MacArthur (Supreme Commander of the Allied Powers—SCAP) was charged by the US President Harry Truman with commencing the process of post-war reconstruction in Japan. One of the Sections of SCAP was the Civil Communications Section (CCS), which included an industrial division that was responsible for working with, and advising, Japanese communications equipment manufacturers (Hopper & Hopper, 2007a).

Homer Sarasohn (*b* 24 May 1916; *d* 28 September 2001) was an American radio engineer. During the Second World War, he fought with the 161st Airborne Engineers, until he was given a medical discharge around Christmas 1943. He proceeded to work on Project Cadillac (to do with radar) at Massachusetts Institute of Technology, and then subsequently put his radar knowledge to civilian use in the development of transcontinental microwave transmitters. His responsibilities included ensuring a rapid transition from prototype to production. He was summoned to Japan (see Appendix A, Item 1) arriving in April 1946, aged 29 years. In Sarasohn's words (Sarasohn, 1997; see Appendix A, Item 2),

[MacArthur] … issued a series of edicts. The first of these were:

- Japan's military forces would be dissolved.
- The *zaibatsu*, the industrial cartels that had supported the military's war adventures, would be abolished. Their executive managers would be removed from positions of influence.

Other proclamations addressed these subjects:

- Workers would be free to form and join labor unions.
- Women would have the same legal status as men.
- Democratic forms of education would be established. Elementary school education would be compulsory.
- Child labor would be banned.
- Political prisoners would be released from jail. The secret police would be abolished.
- Freedom of religion, thought, and political expression would be the right of all people.

But then, a practical problem surfaced. How were the people to be informed about these edicts, their meaning, significance, and all the other details involved? There was a group in SCAP headquarters called the Civil Information & Education Section (CI&E) charged with responsibility for getting the word out to the public. It was important that the people be told about these reforms. They amounted to a major cultural change affecting their lives. The immediate stumbling block, however, was that at this time there were no widely circulated newspapers. The telephone system was not working. It was difficult to travel around the country. There was no radio broadcast system. A communication facility was needed, but none was available.

There was another group in SCAP headquarters working to solve that problem. It was the CCS. Its immediate task was to restore the local and long-distance telephone and telegraph networks to an operating condition. But, the problems being encountered were staggering. Aside from the wartime damage that had been done to the central and switching stations, and the loss of interconnecting lines that had been destroyed, the lack of adequate maintenance work during and prior to the war years resulted in much of the equipment now being unusable. Added to this, there were few trained technicians available to be put to work to correct the situation.

Even more serious was the total lack of a supporting communications equipment manufacturing industry. What was needed was a dependable source of supply of equipment and system components, such as telephone instruments, switches, cables, vacuum tubes, electrical relays, and transformers. These were the kinds of things it would take to restore the networks to operation. But, the companies that had produced these products previously were now, to a large extent, out of business. Their factories had been destroyed. Their workers had been drafted into military service or had otherwise disappeared. Machine tools and production equipment had been lost or had been deployed out to the countryside to escape air-raid damage. An entire class of senior managers had been dismissed in the *zaibatsu* purges. This was the situation the Industry Branch of the Civil Communication Section faced as the Occupation got under way.

I was in charge of the Industry Branch. My assignment was to do mainly three things:

1. Supply domestic radio receivers to the Japanese people as an immediate communications medium in support of SCAP's CI&E Section. (We would use army transmitters as the broadcast source.)
2. Meet the needs of the Occupation Forces (and also domestic users) for a reliable nation-wide telecommunications facility.

3. Assist the Japanese communications equipment manufacturing industry to become a major contributor to a revived national economy.

According to Sarasohn's account, he devoted his efforts to the first of these tasks, with a colleague (Wilbur Magill, from American Telephone and Telegraph (AT&T)'s Hawthorne plant; see Appendix A, Item 3) working on the second. He also commented, "I did not worry too much about the third task. If we were successful in accomplishing the first two, the other would take care of itself". (In fact, Sarasohn related in the interview that a measure of success of their efforts was that this was the first Japanese industry to be taxed, after the war.)

Sarasohn continued:

It was not easy getting to this point. There were physical problems, and there were cultural problems. Among the more pressing physical problems were these. Factory sites had to be cleared of rubble so that shacks could be put up to house production machinery and workers. Machinery had to be installed, repaired and refurbished. Workers had to be recruited and trained. Supplies and raw materials had to be located and brought in. Supervisors and managers had to be chosen, some almost at random, and put in place. Most of them were strangers to their jobs. They came with little or no managerial experience. In their previous positions, they essentially had been conduits for the flow of instructions between their superiors on one side, and the workers on the other. They were not business planners. They were not leaders nor decision makers. They were more accustomed to following orders, rather than giving direction. They had little understanding of industrial strategy or policy. They were confused, lacking in self-confidence and uncomfortable in the positions into which they had been force-fit. They had to be instructed on a day-to-day basis how to set up, run, and manage a mass production system. And, that is what we in CCS did.

Product assignments were given out, manufacturing quotas were set and delivery schedules established. Progress was closely monitored by continuous plant inspections. Direction and assistance were given on site, and on-the-job training was a requirement at every location. I had no illusions as to the level of product quality we would achieve. Pre-war Japan's commercial products were not examples of high quality. In fact, the legend "Made in Japan" stamped on the bottom of an item was a notice to the buyer not to expect a high degree of reliability. This was also true of Japan's war-time production. For example, Koji Kobayashi, who became chief executive officer of Nippon Denki, wrote in an article published in the magazine *Quality Progress* of April 1986, "During the war, NEC manufactured military communications equipment. However, the quality was not good. For example, I remember that the yield of vacuum tubes for aircraft was one percent. We promptly studied the design of experiments and took measures to improve quality, but the prevailing policy was 'one tube today rather than ten tubes tomorrow'".

Quality problems are first and foremost management problems. What better proof of that is needed than Kobayashi's statement? If the leaders of an enterprise do not know and understand that quality is the essence of their business, it is inevitable that they are doomed to failure. And, if they do not know the elements that comprise the system of quality, their fate of failure is sealed.

Quality control is not a "band-aid". It cannot make a bad system good. It is not a therapy to be applied to an ill-conceived or poorly managed function. To be effective as a control, the total process to which it is applied must be well designed to begin with.

By the middle of 1946, a start had been made in reviving the communications equipment manufacturing industry. But, there was such a great distance yet to go. Production facilities were primitive and unreliable. Working conditions were deplorable. Materials wastage was intolerable – all the more unacceptable because raw materials were hard to come by and their cost was exorbitant. Work spaces were contaminated by dust and dirt. In this environment, upgrading product quality was impossible.

The new managers had to be brought to the realization that we were now in the process of building for the future. From now on they would be dealing with the demands of mass production and modern technology. That meant they would have to understand the concept of a total system in which every part was important and interrelated. It was essential to know the functional relationship of each part to its adjacent part. The ordinary workers seemed to have a better understanding of this concept than their managers. What the workers lacked in skill was offset by their industriousness and honest effort. There was no slouching on the job on their part. They seemed dedicated to the idea of making a personal contribution to rebuilding their country and their lives. But, they needed managers who were leaders.

With this in mind, I called a meeting of plant managers at my office in the Dai Ichi [Daiichi Seimei] Building. I had the managers gather around a large table in the conference room. I sat at one end with my interpreter. My agenda was to form a consensus from the suggestions they would volunteer as to the major manufacturing problems they recognized that had to be resolved. I began by saying we had made a pretty good start on the production of radio receivers and their components. Nevertheless, I was still disappointed with the level of quality being achieved. The poor yield was causing an unacceptable waste of valuable materials and worker time and effort. I looked around the table at the men. I asked them to tell me what in their opinion was the reason for the problem, and what action should we take to cure it. My purpose was simply to get them started on some analytical and creative thinking. I wanted to get them out of their old habit of only taking orders from higher authority. I wanted to get them involved in *participative management*.

At first there was dead silence. They seemed shocked and surprised. No one had ever asked for their opinion on anything before. I put my question to them again. Then, they all got up and moved down to the far end of the table. They began a discussion among themselves. This went on for a while and I became increasingly impatient. I turned to my interpreter and asked him what was it they were talking about. Why couldn't they come up with a quick answer to what I thought was a simple question? He said the men were trying to decide upon a response they hoped would be "most pleasant for me to hear". It didn't matter that their answer might not fit the facts. It was more important to them that I not be disturbed. That was not the answer I wanted, and that was not the relationship I wanted with these men.

That episode made me adopt a couple of firm resolutions. First, I would learn as much as I could about Japanese language, culture and mentality so that, in the future, I could deal with the people in a direct and forthright manner without having to depend upon an interpreter. Second, I would break through the tradition that insulates Japanese executives from personal accountability for what happens in their areas of responsibility. Ceremony and circumlocution would be replaced with positive action. My objective was to get these managers to recognize they had serious operating problems that demanded prompt attention. The list was imposing: workplace cleanliness, scheduled machine maintenance, on-time work flow, effective job training, realistic quality standards, and much more. Each of these items called for careful analysis, timely decisions, corrective action and, above all, management follow-through.

That first meeting was followed by a series of other such meetings. Each attendee had to be prepared to identify an operating problem and suggest its solution. Each one was committed to go back to his own company and hold similar discussions with his people. My idea was simply to get everyone involved in an on-going process of *continuous improvement*.

I particularly wanted to inculcate these managers with three fundamental concepts:

Progressive management demands of each person:

* *Commitment* to the defined goals and spirit of the enterprise.
* *A personal sense of Ownership* of and in the organization.
* *Feedback*, up, down and across the lines of the organization, of the information needed to do the job right the first time; of the kind that keeps the sense of commitment and ownership alive and well.

Some Japanese official statistics encompassing the period in question are given as Item 4 in the Appendix.

3.1.2 Union of Japanese Scientists and Engineers (JUSE)

In May 1946, two months after Sarasohn arrived in Japan, the Union of Japanese Scientists and Engineers (JUSE) was created, with founding chairman Ichirō Ishikawa.

3.1.3 Visits to Japan by W. Edwards Deming

There are various reports of W. Edwards Deming visiting Japan in 1946, 1947, and 1948:

* Nancy Mann, in her book *The Keys to Excellence* (1989, p. 14), says that

  ... in 1946 ... Dr. Deming made a trip around the world under the auspices of the Economic and Scientific Section of the U.S. Department of War. While he was in India working with Mahalanobis ... he received instructions to continue on to Japan. He did so and stayed there for two months to assist U.S. occupation forces with studies of nutrition, agricultural production, housing, fisheries, etc. Thus, he became friends with some of the great Japanese statisticians.

* In an article on the website of the Japanese Statistics Bureau, Kitada (1995) recorded that "In response to the request of the General Headquarters to the U.S. Government on October 31, 1946, the first mission came to Japan on December 22 of the same year, just before the Statistics Commission was organized, to research and recommend on reform of statistical affairs of the Japanese Government and other matters. The mission conducted research in cooperation with the Statistics Commission and, on May 28, 1947, submitted to the General Headquarters a recommendation titled 'Modernization of Japan's Statistics'. The leader of the mission was Dr. S. A. Rice, Assistant Director of the Bureau of Budget (BoB) and Director of the Statistical Standards Department of BoB of the Executive Office of the President ... The mission also included ... Mr. W. E. Deming. Dr. Rice submitted [a report] to

the General Headquarters on January 11, 1947 and returned to the U.S. at the end of the month. [At the time, Rice was also President of the International Statistical Institute—NIF.]"

In the Manuscript Division of the Library of Congress, where Deming's records are stored, the *International File, 1930–1992, n.d.* reports that "In 1947, Deming visited Japan as a statistical advisor to the Supreme Command of the Allied Powers". This is supported by a statement in a letter from Sarasohn to Lloyd Dobyns (quoted more fully below) to the effect that "Deming was a statistician who had been on loan to SCAP from the Census Bureau in 1947 when we were trying to understand the local demographics relative to food distribution, health statistics, social services, etc". Deming may also have been doing some planning for the 1951 census in Japan. Various reports suggest that this trip occurred in January 1947, as part of a world trip that also included a visit to India in December 1946–January 1947.

The import of these two paragraphs is that Deming may not have joined Rice's mission in Japan until January 1947. Deming is also recorded as giving a lecture on sampling at the Institute of Statistics and Mathematics at Tokyo in March, 1947.

- Mann (1989, p. 15) also quotes Deming as making a visit to Japan in 1948 "… to do more of what I had done before".

Myron Tribus (1994) records that "… when General MacArthur needed to make a population survey in Japan in 1948, he called upon W. Edwards Deming", although he does not state that Deming actually made the visit in 1948. Deming visited Japan in 1950 for this purpose (amongst others).

3.2   Interpretation

There is little of contention about this period. Sarasohn was working on the basic task of developing manufacturing capability for domestic radio receivers, JUSE had just been formed, and Deming was involved in census work and other basic studies relating to restoring the Japanese economy.

4   1948–1950

4.1   Data

4.1.1   CCS activities

In November 1948, Charles Protzman arrived in Japan to join CCS. Kenneth Hopper (1982, pp. 15, 19) provides the following background information:

> After joining Western Electric in 1922, Protzman had risen to senior production management levels before being assigned to SCAP at the age of 48 as a civilian advisor to the Japanese communications industry.

... Protzman had been instructed before he left the U.S. that, with his substantial knowledge of manufacturing, one of his prime duties would be to help improve quality. When he arrived in Japan in 1948, he concluded that while "in some individual cases, good quality levels had been attained", in general "it was still far below reasonable standards".

[Protzman continued] "I understood my job was to advise the Japanese on rebuilding their communications system. I found, however, that they did not understand and apply the systems and routines of production management. Within a month of arriving in Japan, I had concluded that rather than try to correct each company individually, we should present a set of seminars on the principles of industrial management for top company executives. I recommended this, and found Sarasohn in agreement".

Hopper (1982) recounts that there was resistance to this from their immediate superior officer, so the preparatory work was done discreetly. However, this superior officer was replaced by Frank Polkinghorn (Appendix A, Item 5), who was far more supportive, and they were able to proceed. Continuing with Sarasohn's (1997) account:

In carrying out our "arms length" posture with the communications industry, we in CCS adopted two new concepts. One was product quality certification. The other was management qualification. My first move was to establish a national electrical testing laboratory. Managers and engineers cooperated with me in drafting and agreeing to uphold performance specifications and test criteria that covered the entire spectrum of communications products. An edict was issued that required all electronic, radio, telephone, telegraph and related equipment to be type-tested and quality certified by this laboratory before being offered to the public. If approved, all production units must then adhere to the same test criteria. To ensure continuing compliance by the manufacturers, tests would subsequently be run from time-to-time on items taken at random from store shelves. If there were any failures, manufacturers would be required to withdraw all products of that type until a re-certification test was completed.

The rationale behind this was simple. By this time, radio receiver production was meeting our goals. The CI&E program was getting through to the public. The telecommunication system was working with an acceptable degree of reliability. Now, we could go back and make each manager individually responsible for the quality of his product and his function. Manufacturing quantity might suffer. But, the long-term benefit of quality control would make that cost quite acceptable.

(In other words, Sarasohn established an electrical standards testing laboratory, the fore-runner of the Electrical Test Laboratory that still operates in Japan.)

> The second step taken toward the end of 1949 was aimed at improving and broadening the quality of management. Up to this time, junior level managers had been squeezed into senior level positions. By and large, they had responded admirably to the challenge. They were becoming increasingly effective. Nevertheless, it was obvious there was no depth to the available resource. Moreover, the cultural influence of the feudal environment from which they had emerged was still quite evident. It was clear to us that an intensive management training course was needed.

> To get a more precise measure of just what the scope and content of that course should be, a colleague of mine and I made a detailed investigation of six companies typical of all those in the communications industry. What we found was disturbing. In the report written at the conclusion of our survey, we stated, in spite of the progress being made at the factory level, it was clear that "the weaknesses of management at the top level were causing a tide of regression which, if allowed to go unchecked, might well culminate in the collapse of the industry".

> We said these top level executives had to come to a management school. CCS would be that school. Charles Protzman, my colleague who was an AT&T industrial engineer, and I would be the teachers. We knew there was no textbook available that covered the subjects we had in mind for our students, so we would write the textbook ourselves. Certain rules would apply to our school:

> * We would select the senior executives who would be our students. They would be required to attend. Substitutes would not be permitted. Certain government officials and university professors would also be selected to attend.
> * Classes would meet four days a week for eight consecutive weeks, four hours a day. There would be homework for our students to do [cf. Item 6].
> * Each student would be expected to apply each lesson learned to his company as soon as possible.
> * The CCS Seminar lessons would be repeated in each student's company. The students of this first course would be the teachers of the next level of managers.
> * The final examination for each of our students would be the progress made in his company in one year's time.

Protzman and Sarasohn were now in a position to propose teaching the Japanese about industrial management. However, opposition emerged from the Economic and Scientific Section (ESS), which was essentially responsible for all Japanese industry except communications. ESS officers were concerned about the impact on American companies if Sarasohn and Protzman proceeded with their course.

> When our plans for the CCS Seminar became known elsewhere in SCAP headquarters, a series of objections were raised. Statements were made that we should not teach the Japanese about progressive management; there was a competitive danger in raising the industry's productivity level too high; we might make it more difficult for American companies to get a commercial foothold in Japan.

> We were not dissuaded by such argument. So, the matter was brought to General MacArthur's attention for his final decision.

> The spokesman for the opposition and I went to MacArthur's office. He made his presentation first, pointing out all the reasons why the idea of the CCS Seminar was so bad. Then I got up and spoke for my allotted time—fifteen or twenty minutes. My main point was that strong managerial leadership built upon the base of the country's industrious workers would assure a progressive future for Japan. During all this time, MacArthur sat at his desk smoking his corncob pipe, saying not a word, never changing the expression on his face. "I finished my presentation and sat down, thinking that I had failed to get my story across to him. Suddenly, he got up, and started walking toward the door". He stopped, turned around and glared at me. "Go do it!" he blurted; turned around, and walked out.

> (Sarasohn, 1997)

As recounted by Sarasohn in the interview with Myron Tribus (and reported in other sources, e.g. Dobyns and Crawford-Mason, 1991), he and Protzman then went off to a hotel in Osaka for one month, sat down in separate rooms, and proceeded to write a complete manual on Industrial Management (Sarasohn and Protzman, 1949). Sarasohn (1997) continues:

> It is not a philosophical or academic treatise. It lays a practical and pragmatic foundation for progressive management. Protzman's half of the book covers such subjects as manufacturing engineering, cost control, factory layout and inventory management. My half deals with management policy formation, long range strategy and planning, organizational structures, research and product development and quality

control. Statistical quality and process control occupied more space in the book and more time in the lectures than any other subject.

The first CCS Management Seminar was then presented during the 8-week period September 26–November 18, 1949. It was for top management only; no substitutes were permitted. As planned, the course ran for 4 hours each day, 4 days a week, with homework each night. As Sarasohn relates, there was no final examination at the end of the course. Rather, participants were told that their success or failure would be judged by the performance of their companies at the end of 12 months. A photograph of the participants may be seen in Figure 1. The list of participants included Takeo Kato from Mitsubishi Electric, Hanzou Omi from Fujitsu, and similar top executives from Furukawa, Hitachi, N.E.C. and Toshiba, or their predecessor companies.

The second CCS Management Seminar ran in Osaka from November 21, 1949 to January 20, 1950, and included Bunzaemon Inoue from Sumitomo Electric, Masaharu Matsushita from Matsushita Electric, and the top executives from Sanyo Sharp (or their predecessor companies). Akio Morita and Masaru Ibuka, the founders of Sony Corporation, were schooled separately by Sarasohn (Donkin, 2001, Chapter 15). Polkinghorn wrote the Preface for the Second Seminar; see Hopper & Hopper (2007b).

Polkinghorn introduced each Seminar, and the presenters were Sarasohn and Protzman (with Sarasohn teaching in Japanese). In particular, Sarasohn presented the six sessions on Quality Control (which was allocated more time than any other topic).

It is worthwhile to look at how Sarasohn commenced the Seminars, as it shows the primacy given to the issue of Quality (Sarasohn & Protzman, 1949, vii–viii).

> Why does any company exist? What is the reason for being of any business enterprise? Many people would probably answer these questions by saying that the purpose of a company is to make a profit.
>
> In fact, if I were to ask you to write down right now the principal reason why your companies are in business, I suppose that most of the answers would be something of this sort.
>
> But such a statement is not a complete idea, nor is it a satisfactory answer because it does not clearly state the objective of the company, the principal goal that the company management is to strive for. A company's objective should be stated in a way which will not permit of any uncertainty as to its real fundamental purpose. For example, there are two ways of looking at that statement about profit. One is to make the product for a cost that is less than the price at which it is to be sold. The other is to sell the product for a price higher than it costs to make.

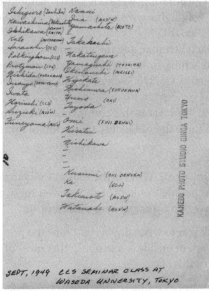

Figure 1. Participants in the first CCS Seminar Class, photographed at Waseda University, Tokyo in September 1949. The reverse side of the photograph identifies the participants. Photo kindly supplied by Lisa Sarasohn.

These two views are almost the same—but not quite. The first implies a cost-conscious attitude on the part of the company. The second seems to say whatever the product costs, it will be sold at a higher price.

There is another fault that I would find in such a statement. It is entirely selfish and one-sided. It ignores entirely the sociologic aspects that should be a part of a company's thinking. The business enterprise must be founded upon a sense of responsibility to the public and to its employees. Service to its customers, the wellbeing of its employees, good citizenship in the communities in which it operates—these are cardinal principles fundamental to any business. They provide the platform upon which a profitable company is built.

The founder of the Newport News Shipbuilding and Dry Dock Company, when he was starting his company many years ago, wrote down his idea of the objective—the purpose—of the enterprise.

He put it this way. "We shall build good ships here; at a profit if we can—at a loss if we must—but, always good ships".

This is the guiding principle of this company and its fundamental policy. And it is a good one too because in a very few words it tells the whole reason for the existence of the enterprise. And yet inherent in these few words there is a wealth of meaning. The determination to put quality ahead of profit. A promise to stay in business in spite of adversity. A determination to find the best production methods.

Every business enterprise should have as its very basic policy a simple clear statement, something of this nature, which will set forth its reason for being. In fact, it is imperative that it should have such a fundamental pronouncement because there are some very definite and important uses to which it can be put. The most important use of basic policy is to aim the entire resources and efforts of the company toward a well defined target. In a general way it charts the course that the activity of the company will follow in going toward the target. (See Fig. III.)

Making a clear statement of the objective of the enterprise is like providing a target for a man shooting an arrow with a bow. Figure No. III shows such a man who represents company management holding a bow which represents company policies and an arrow which represents the total efforts and resources of the company. If no target is provided for management (the man), toward which company efforts and resources (the arrow) can be aimed and directed, company policies (the bow), no matter how good they may be will be utterly useless. But altogether, policies, efforts and resources and ultimate purpose to which they are to be put are all part of a single picture. Any one part has a definite intimate inter-relationship with every other part, and no one part is able to stand alone. Each demands the co-existence of the other elements in order to comprise the total picture which is the entire business enterprise.

A great advantage to be gained in a statement of the objective is the stabilizing effect it is bound to have on all features of the organization. For one thing, employees will better understand the use to which their efforts are being put in relation to the total enterprise. At the same time, the statement of the objective will build up confidence in the customers by letting them know just what they can expect from the company. Then too, the part that everyone in the company must play in relation to the attainment of the desired goal is more easily recognized because of the stated objective.

The statement of purpose also serves as a point of measurement, a standard, against which current operations can be measured in order for management to assess the accomplishments of the company. Not the least of the advantages to be gained is the opportunity afforded for pre-determining all the factors involved in attaining the sought-for goal.

Thus, the basis of the organization lies first of all in the enunciation of the basic policy, the fundamental objective of the enterprise.

Obviously, there can be as many different stated objectives as there are different business enterprises. But one point stands out clearly. A very necessary preliminary to the establishment of any company is a clear, concise, complete statement of the purpose of the company's existence.

After presenting these two courses, Sarasohn and Protzman had planned to continue with a program for middle and plant level managers (see Appendix A, Item 7):

Our follow-on plan was to continue after the CCS Seminar with a series of shorter, more detailed courses (see Appendix A, Item 8) aimed at middle and plant level managers. The topics to be covered included industrial engineering, manufacturing cost control, product development transition, and statistical quality control. The first of these was presented in Tokyo in the Spring of 1950. However, our plan was suddenly interrupted when South Korea was invaded by forces that came down from the north. The focus of our attention in SCAP immediately shifted to that crisis.

Nevertheless, it was important to me that we not lose the momentum toward success that had been building over the past several years. At least, I wanted our work in quality control to carry on even if we could not continue it ourselves. I tried to get Walter Shewhart to come to Japan to be the teacher. But, he was ill at that time, and was not available. We thought of others who might take over. We finally decided Dr. W. Edwards Deming should be invited. He was a statistician and an early student of Shewhart who is deservedly known as the "Father of Quality Control". Deming came and was very well received. His contributions to the improvement of quality management made a lasting impression upon the Japanese industrial scene.

(Sarasohn, 1997)

In fact, Sarasohn had gone so far as to prepare a book on Statistical Quality Control in Japanese (Sarasohn, 1952), and may have taught from this during return visits to Japan.

And so Protzman and Sarasohn departed, Protzman returning to the USA and Sarasohn accompanying MacArthur to Korea. Protzman and Sarasohn each received letters from the officer in charge of CCS, acknowledging their contributions. Sarasohn's letter (Back, 1950a) read:

GENERAL HEADQUARTERS
SUPREME COMMANDER FOR THE ALLIED FORCES
Civil Communications Section APO 500

8 August 1950

Dear Mr Sarasohn,

On the occasion of your departure from Japan, I wish to express my appreciation for the invaluable contribution you have made to the Occupation in your capacity as Industrial Engineer in the Research and Development Division of the Civil Communications Section, General Headquarters, Supreme Commander for the Allied Powers.

Your outstanding work in connection with the rehabilitation and re-orientation of the Japanese communications equipment manufacturing industry has materially aided in the task of re-establishing Japan as a completely stable nation.

Through the introduction of the ideas of scientific industrial management you have helped to raise the engineering standards and promote the use of modern manufacturing methods and practices to such an extent that the industry is now one of the most reliable and important adjuncts of the Japanese economy.

I wish to commend you for your outstanding work in introducing in Japan the modern concepts and practices of statistical quality control and for your development of the series of courses in modern industrial management methods through which you have so ably indoctrinated Japanese management personnel in up-to-date techniques of production management, manufacturing methods, production engineering and design engineering practices.

Your advice and assistance on the many problems in the field of research and development in the Japanese communications equipment manufacturing industry have contributed materially to the accomplishment of the Civil Communications Section's mission.

Please accept my best wishes for continued success in connection with your future endeavors.

Sincerely,

GEORGE I. BACK
Brigadier General,
USA Chief, Civil Communications Section

Back's letter to Protzman (Back, 1950b) was dated March 28, 1950 and read:

Dear Mr Protzman,

On the eve of your departure from Tokyo, may I take the opportunity of expressing to you my appreciation for the assistance you have rendered during the period of your assignment to Civil Communications Section, General Headquarters, Tokyo, Japan.

From the date of your assignment as Wire Equipment Engineering Supervisor on 13 November 1948, and later as Research and Development Engineer (Telephone and Telegraph Manufacturing), you have discharged your duties in an exemplary fashion.

Your recent accomplishment in planning, preparing and giving the Management Training courses to the Japanese telecommunications manufacturers have demonstrated your unusual ability to analyze and solve the unique telecommunications problems facing us in Japan. Your achievements in this field may well set the pattern for a truly democratic system of management and may prove the difference between success or failure of the telecommunications manufacturers. Furthermore, your advice and guidance on numerous and varied problems has contributed materially to the accomplishment of Civil Communications Section's mission.

Please accept my best wishes for continued success and happiness.

4.1.2 JUSE activities

In March 1948, JUSE established committee to study SQC. Tsutsui (1996, pp. 305–306) cites sources documenting

… a generous grant [awarded in 1949] from the Economic Stabilization Board [Keizai Antei Honbu] to produce a report on recent technological advances abroad. The project not only ensured solvency, but it allowed the organization's staff to investigate the relevance of new scientific discoveries to Japan's economic reconstruction. … After combing the Occupation's American library at Hibiya and evaluating subjects such as atomic energy and ultra-high-frequency communications, JUSE's leaders finally settled on a topic that could serve as the fulcrum of the organization's research, educational and promotional activities. The new technology judged most relevant and promising for JUSE-sponsored introduction was statistical quality control.

By September 1949, the Quality Control Research Group was in a position to present a "Basic Course on the fundamentals of statistical quality control". It was repeated in 1950.

4.1.3 Deming's activities

The background to W. Edwards Deming's return visits to Japan is a matter of speculation, and so is taken up in Section 4.2 below.

Professor Tadashi Yoshizawa has kindly provided the following information from Moriguti's (1987) paper (in Japanese):

> Deming visited Japan [in 1947 and 1948?] on a task related to official statistics. During his stay, he had a meeting with Japanese academic professors and statisticians and gave his books and new information on statistical studies. Professor Jiro Yamauchi (University of Tokyo, at that time) attended the meeting. He suggested subsequently that Professor Sigeiti Moriguti read one of Deming's books: "Statistical Adjustment of Data" (Wiley, 1943). Moriguti began to translate the book and to correspond with Deming to clarify questions concerning translation of the book. In 1949, Moriguti was informed by Dr. Deming of his plan to visit to Japan in 1950 in their exchange of letters concerning the translation. Then Mr. Koyanagi of JUSE learned of Deming's intention from Moriguti and wrote a letter to Dr. Deming asking Deming to present an SQC Seminar at JUSE. The translation was published in August, 1950.

Mann (1989, p. 15) quotes Deming as saying that

> In 1948, I went again to Japan, this time for the Department of Defense, to do more of what I had done before.

Of greater significance was the trip in 1950 when, during 10–18 July, Deming taught a course on "Elementary Principles of the Statistical Control of Quality" (Deming, 1950) to an audience of Japanese engineers and technicians.

On the same visit, he had dinner with the presidents and senior officials of some of Japan's leading industries to talk about quality (Dobyns & Crawford-Mason, 1991, Chapter 1).

This trip led to many other trips in subsequent years, with Deming making presentations attended by all levels of management. He donated the royalties from his notes, published in English and Japanese, to JUSE, and they were used to set up the Deming Prize. He also recommended to the Japanese that they study Joseph Juran's (1951) *Quality Control Handbook*. Juran's first visit occurred in 1954, the beginning of his own contribution to Japanese management.

4.2 Interpretation

This section addresses two issues:

1. The "scholarly authority" of Sarasohn and Protzman in preparing the CCS Management Seminar.
2. The origins of the invitation to Deming to visit in 1950 for the purpose of teaching Statistical Quality Control.

It is difficult to find agreement among the many competing accounts, especially in relation to (2). The difficulties include:

- Up until the late 1980s, Deming himself professed to be unaware of any contributions by other Americans prior to his lectures in 1950 (see Appendix A, Item 9). On p. 486 of the 1986 edition of *Out of the Crisis*, Deming states:

   The whole word is familiar with the miracle of Japan, and knows that the miracle started off with a concussion in 1950.

John Butman (1997, p. 96) also relates an incident described by Dobyns & Crawford-Mason in which Deming, at a dinner party in the mid 1980s, was

   [p]ressed to say what had made the difference in Japan, a question he had always avoided ... Deming drew himself to his full seated height, slapped his hand on the table, and said firmly and finally, "one lone man with profound knowledge"—referring, of course, to himself.

- The point made by several different people: How could Sarasohn, with no proper management experience (especially senior management experience), be in a position to write and deliver a course on management? (Protzman had had several years of experience in senior production management.) In the interview that Myron Tribus conducted with Sarasohn, and subsequently in a letter to Homer Sarasohn dated 20 July 1988, Tribus captures the essence of the issue:

   I for one still continue to ponder how it came about that a youngster, such as you were, had the audacity to impose upon the Japanese, with the force of the military behind him, a method of management which was NOT the one that won the war, was not the dominant mode of US management, was not born of personal experience in managing a large enterprise and was not just lifted from the textbook of some acknowledged leader in management. Instead the philosophy represented ideas born out of a sensitivity for humanism, out of an engineer's logic and a feeling of what is "right". What you had to say represents, today, 43 years later, what we now regard as the best philosophy of management—one which excites the imagination and spirit of thousands of people. I know this to be true, because it forms the basis for what I teach. Using these ideas I have lectured on 5 continents and found the same eager reception. What is the genesis? Who put it all together for the first time? Was it really Sarasohn? If so, how did it happen??

The answer to (1) can be found in Sarasohn's indirect response to this, which is contained in a letter to Clare Crawford-Mason two years later, dated 14 August, 1990. Sarasohn quotes the above passage from Tribus' letter, and then continues:

It really was Sarasohn who put it all together. It happened because there was a special situation and a special need. The situation was the American Government's decision to rebuild the Japanese economy; and the need was to start the process, literally, from the ground up. There was no precedent to follow, and there were few, if any, material or human Japanese resources available to begin with. So, I was not bound by any American tradition, and I had free rein to do what had to be done.

You asked whence came my authority. Authority stems from two sources. One is implicit in the job responsibilities that are assigned. The other is given by the people who willingly follow the leader. My authority came from MacArthur and from the Japanese people who followed my lead. And, I made the best use of it.

I was no "youngster" at the time. I was almost thirty when I arrived in Japan. I did what had to be done. And, by means of logical analysis, decision and determination, the job was accomplished. I am puzzled that there should be any wonderment about that. By the time Deming arrived on the scene to make his contribution (which I do not minimize), a solid industrial operating base had been laid. He carried on from there.

There are two reasons why I believe it is reasonable to accept Sarasohn's explanation.

The first is that it makes sense. Sarasohn was well read, a student of history, and a particular admirer of the achievements of the British in their Industrial Revolution. His engineering experience meant that defining a goal, developing a plan, executing the plan, and delivering "on time, to specification and within budget", so to speak, were second nature to him. In fact, major advances in the success of Quality Management have the names of engineers associated with them; for example William Conway at Nashua Corporation, the first leader of a Fortune 500 company to find an effective way of putting Deming's management philosophy into practice through his concept of Waste (e.g. Conway, 1992); and Myron Tribus, not least through his re-expression in plain language of the key elements of Quality Management that now underpin the Baldrige Business Excellence framework. Sarasohn combined his engineering know-how with knowledge extracted from his learning; all informed, in Tribus' words, with "a sensitivity for humanism", to create and implement an outstanding methodology. In Sarasohn's words, "I did what had to be done". Four decades later, he was still surprised that "that there should be any wonderment about that".

The second derives from personal acquaintance with Sarasohn. It is evident from the interview that Myron Tribus conducted with Sarasohn in 1988 that he accepted Sarasohn's account completely. David Howard, who got to know him through a joint consulting project (e.g. Howard, 2008) wrote, in a personal communication, that

> To the extent that Homer was modest beyond measure – a quality not so evident in the members of "his" modern-day peer group – I would tend to weigh my views in his direction while remembering – as he would insist – that variation infests all things, including our actions and memories.
>
> To me the thing that matters is that his efforts, together with those of a few other Americans, helped Japan make the grade after the War. Homer was perhaps the most distinctive, yet self-effacing, of those. I always thank him when I look at my Sony television.

I (NIF) had the great good fortune to get to know him well, late in his life, when he entrusted me with his only copy of his CCS notes (unopened in nearly 50 years) to re-publish. I found him to be a person of total integrity, with a dry, self-deprecating sense of humor, and without pretense. He was satisfied that he'd carried out his tasks well, made no false claims about his achievements, and gave credit where it was due. In our discussions, he expressed puzzlement that Deming had been unaware of CCS activities until the last few years of his (Deming's) life, but did say that he'd received a letter from Deming around 1990 acknowledging that CCS had made a contribution. (Sarasohn was unable to find the letter when he mentioned it, and no copy has since come to light.)

The answer to (2) leads into rather more controversial territory.

Tsutsui (1996) has sought to re-evaluate Deming's achievements in Japan, and those of JUSE, drawing extensively on Japanese resources. However, the analysis appears not to have taken account of a wealth of American resources (not least, materials about Deming's other accomplishments: see Appendix A, Item 10), nor is there reference to any attempt at obtaining comment from Sarasohn. As a consequence, little weight can be ascribed to its conclusions, which are best characterized as partial.

A sounder basis for understanding what took place can be found in the scholarly writings of Myron Tribus, whose views about Deming's contributions are summarized in a memorial article, Tribus (1994).

Another source has also become available, namely Sarasohn's own account, in response to a specific query from Lloyd Dobyns. Referring to an earlier letter from Sarasohn, Dobyns writes in a letter to Sarasohn dated 8 October 1990:

> In your memo, you say you "arranged his (Deming's) invitation to come…" to Japan. In the interview [Tribus' interview with Sarasohn] you said you first wanted Walter Shewhart, then asked ESS to get Deming. In an article written in 1960 Kenichi Koyanagi, former managing director of JUSE, reprints the letter he wrote Deming in March, 1950, inviting him to give a series of lectures for JUSE, and he reprints Deming's letter from April, 1950, accepting. In the letter, Deming offers to lecture two to four hours a day, then he adds, "However, I should explain before being too definite that my time will be under the direction of Mr Kenneth Morrow, Research and Programs Division, Economic and Scientific Section, SCAP". Koyanagi in 1960 wrote that Deming did sampling for SCAP in 1947 and '50. Deming has told me that he was invited to lecture by JUSE and Japan's leading industrialists. I have a copy of "Japan 1950", the privately-published diary he kept of the trip, that says the same things.

> In the same article Koyanagi says he learned of Deming's planned visit in 1950 from Dr. Shigeiti Moriguti. He also wrote, "The Civil Communications Section, SCAP, urged Japanese communication equipment makers to adopt quality control methods, offering educational service for this purpose". He mentions the "Japanese Management Association and some other private organizations" that also helped, then he adds "independent from these organizations" JUSE started working on SQC.

> I think you can understand my confusion. Did CCS and ESS coordinate with JUSE or, perhaps, suggest Koyanagi's letter to Deming. Or was Moriguti an emissary from SCAP? I can see several ways that SCAP could have "arranged" for Deming to give the lecture series, but I can't find in my research how it was done. Any help you can give me on this I will sincerely appreciate.

[The Associate Editor of ISR noted that Sigeiti Moriguti was then Associate Professor and later Professor, finally Emeritus Professor at the University of Tokyo, and ISI President in 1987 at the time of ISI Session in Tokyo.]

Sarasohn responded three days later (11 October 1990):

Dear Mr Dobyns,

Thank you for your October 8 letter. I particularly appreciate your request for clarification of a couple of salient points of history. I will try to respond carefully, objectively, and as accurately as my memory of the events I lived through in Japan will allow.

It will come as no news to you, I am sure, that people—deliberately, or unwittingly—re-shape history as they wish it had been, or to support a point of view they profess now, or to adorn themselves belatedly with "crowns of ivy". To some extent, this is what is happening in Japan now, and what has happened in the case of the two points about which you ask.

[Material about the first query omitted]

Walter Shewhart was an expert on quality control in manufacturing and in engineering. Deming was a statistician who had been on loan to SCAP from the Census Bureau in 1947 when we were trying to understand the local demographics relative to food distribution, health statistics, social services, etc. He branched out later to the industrial application of statistics.

After the CCS Management Seminars were established and thriving, I wanted to have a specialized, concentrated course on quality control methods specifically for plant managers as a follow-on to the quality concepts, philosophy and policy issues I had dwelled on with the senior executives who were my seminar "students". I did not want these people to be fixated on the mechanics of statistics. Rather, it was essential that they understand the entire management function and all of its related parts as a SYSTEM, including the component that was statistical analysis. In other words, statistics was merely a tool that is used to gain an ultimate objective. It is not an end in itself. I felt, and feel, strongly on this, and it has put me at odds with some other folks who speak on the subject of Quality Control.

For example, I had to put my foot down unceremoniously with Koyanagi, Koga, Ishikawa and some others of the Union of Japanese Scientists and Engineers (JUSE). They had a simple-minded view. They had come across some early AT&T reports. It occurred to them that all one had to know was the mathematics of statistics—that was what enabled the United States to win the war!! They saw quality control as an academic exercise. Fortunately, there were others, such as Nishibori and Mabuchi, who were level-headed and wanted not only to learn, but also to understand. So, I blocked the JUSE effort to go wandering off on the wrong track. At the same time, I had another motive. I wanted the plant managers' attention to be focused on the production matters at hand. I did not want their concentration diverted to abstractions they were not yet prepared to handle. It was a question of priorities, and JUSE was off-base.

By 1950 the CCS Management Seminars were off to a good start and the time was now ripe for the detailed course on statistical quality control that I had had in mind. (Incidentally, I went over all of this history with the JUSE people at a banquet they held for me when I was back in Tokyo several years ago.) Koyanagi knew of my plan, and I have copies of his and Deming's 1950 letters. In my view Koyanagi was precipitous, and Deming's reply was appropriate.

It was clear that Shewhart was the right man for the job I wanted done. However, it turned out he was not available. A second choice had to be selected from among other potential candidates. But, another consideration was to open up the follow-on course to, not only the communications and electronics manufacturers, but to the other industries under SCAP control as well. So, I went to my friends in ESS, Ken Morrow and others, with the suggestion that we sponsor the course jointly. Seeing that the CCS Seminars were successful, they agreed. We began to explore who we might get as the instructor. And, we formed a working committee among the Japanese to handle the matter of attendees and other "housekeeping" chores. Koyanagi was a member of this committee.

Remembering his earlier exposure to Japan, and knowing something about his subsequent university occupation (at Columbia? New York University?), I suggested that Deming be given the job. Then I asked the ESS to take the lead on this program because I was preoccupied with establishing the Electrical Test Laboratory as the authority to perform the required quality certification prototype tests of all electronic and communications products that were proposed to be marketed.

I have known that Koyanagi took great personal pride in associating himself with the honor that was reflected from the popular acceptance accorded Deming by the Japanese. It did not bother me then, nor does it now, that he has represented himself as the instigator of Japan's mastery of the quality control effort. Perhaps he needed that to gain stature among his peers.

What does bother me is that he, in his 1960 article, used a few obscure references to gloss over the significant contributions CCS made to modernize Japan's management methods (cf., "… urged… makers to adopt quality control methods, offering educational service for this purpose"). The "help" given by the Japan Management Association was merely administrative. JUSE's work on SQC came later. The fact is, little is known in the United States, and little is admitted in Japan as to the contributions made by the Americans in CCS to the reconstruction of Japan's post-war industry prior to Deming's arrival on the scene. Hopefully, your narration, the television documentary and your book will help to set the record straight.

…

So, to summarize this retrospective, Deming was my second choice. I suggested that ESS take over the program so that there would be a wider application of the quality control commitment in Japanese industry. They agreed to bring Deming over and get him started. There was no "coordination" with JUSE, but they were informed, as were others in the government and associations, of what we were doing. Moriguchi was helpful in this regard. I believe that Koyanagi took it upon himself to write his letter to Deming. I believe, further, that ESS's leadership in the program was less than assertive for two reasons: (1) they were lukewarm players to begin with because of the earlier controversy with CCS, and (2) with the prospect of the Korean war, SCAP's focus shifted across the Sea of Japan.

5 Conclusions

(a) Homer Sarasohn and Charles Protzman taught a number of senior Japanese people how to manage a company, using a system of management (philosophy, structure, and implementation plan) devised largely by Sarasohn and viewed by a scholar of the eminence of Myron Tribus as being consistent with best current management practice at least 30 years later. (For a contrary view, see Appendix A, Item 11.) A very plausible explanation for how Sarasohn, despite his lack of background in line management, was able to devise and deliver the material is that he was an exceptional individual who combined a scholarly historical bent with practical engineering experience in taking products from prototype to manufacture, and was able to bring all of this to bear to produce a practical system of management. *And it worked*, as judged by the progress made by the companies taught by Sarasohn and Protzman and the early recognition of this through early taxation of the telecommunications industry.

(b) Sarasohn had a very general view that Quality is the primary responsibility of management, and that unless this is accepted and adopted by management it is a waste of time introducing technical methods of statistical quality control for operational activities (see Appendix A, Item 12). Whilst he didn't coin the term "Total Quality Management" it captured his approach. In a letter to Malcolm Trevor dated 25 July, 1988, he wrote:

> Matsushita was one of … [the] companies that I started or rehabilitated in the communications industry in that post-World War II period. Karatsu, Kayano, Matsushita (elder and younger) are among the people I worked with. What I tried to convey to them and the others … what is essential in the industrial environment is the quality of management in each of its activities and in the totality of all of its operations. QC is related not merely to product manufacture or to incoming inspection, for example. It is a guiding state of mind, a devotion and dedication. It is a phenomenon, not of statistics, but rather of an integrated system composed of interacting and mutually dependent parts.

(Sarasohn's section of the course notes also reveal very early on that he had a strong stakeholder perspective. The very first section of the CCS manual talks about

> … the sociologic aspects that should be a part of a company's thinking. The business enterprise must be founded upon a sense of responsibility to the public and to its employees. Service to its customers, the well-being of its employees, good citizenship in the communities in which

it operates—these are cardinal principles fundamental to any business. They provide the platform upon which a profitable company is built.

This prefigures approaches half a century later, on drivers of shareholder value, and systemic approaches to performance measurement.)

(a) The approach taught by Sarasohn and Protzman was a development from the work of Frederick Winslow Taylor (1911) on *scientific management*, with critical differences, most notably in its attitude toward the people working in a company. Whereas Taylor's focus was almost entirely on improving efficiency of operations to optimize financial gain, Sarasohn and Protzman recognized in the very first Section of their Manual the importance of "…the wellbeing of its employees…" in building a profitable enterprise.

(b) The remarkable impact of the work of the CCS as viewed by Japanese people involved in the original courses is summarized by Hopper (1982, p. 29). (A 4-day version of the CCS course continued to run until 1974, under the auspices of the Japanese Industrial and Vocational Training Association, and still occupied the number one position of honor in JIVTA's annual catalogue as late as 1982.)

(c) More generally, Hopper (1982) describes how the Japanese themselves evolved scientific management taught by the Americans into Japanese industrial management.

(d) Japan was fortunate to have the opportunity to learn from the collective wisdom of people like Sarasohn, Protzman, Deming, and subsequently others, and then to have leaders in their own community able to develop and implement their ideas and methods so well in the Japanese setting. Adams & Moranti (2007) expressed the nature of the impact as follows: "The CCS offered a solid institutional foundation that Japanese managers adapted to local economic and cultural circumstances, contributing to Japan's spectacular takeoff in global electronic markets beginning in the 1960s".

Finally, a broader observation based not only on the events of this period, but on subsequent developments in America: whilst W. Edwards Deming eventually developed a "System of Profound Knowledge" for management, it required a "System of Profound Knowhow" to bring the system to life in an organization and have it succeed. The latter system was something Deming did not create: his background was as a statistical scientist, with no experience in line management. It required people who had either had line management experience or the sort of disciplined development and production programs typically experienced by engineers (Sarasohn, Protzman, Juran, Tribus, Conway…) to make Quality Management work in practice.

6   Key Sources of Information

The preparation of the paper drew significantly on articles and correspondence recently made available by the late Homer Sarasohn's daughter, Lisa Sarasohn, through the memorial website http:// www.honoringhomer.net. Sarasohn's papers have been catalogued and are housed by the Library of Congress.

Kenneth and William Hopper's website (Hopper & Hopper 2007b) is also an invaluable resource. Other helpful materials include:

- "How Quality First Came to Japan", a videotape made by Myron Tribus of an interview he conducted with Homer Sarasohn on 17 July 1988 at Sarasohn's home in Scottsdale, Arizona. (This interview has been transferred to digital format, with the hope that it will become available via the Web.)
- Sarasohn's own account of this period, presented at a conference in Sydney, April 2007 (Sarasohn, 1997).
- Extensive research by Kenneth Hopper, much of which is summarized or referenced in three important publications: the article "Creating Japan's New Industrial Management: The Americans as Teachers" (Hopper, 1982), based on a 7-year correspondence between Hopper and Bunzaemon Inoue; a subsequent article "Quality, Japan, and the US: The First Chapter" (Hopper, 1985); and a recent book, *The Puritan Gift* (Hopper & Hopper, 2007a) written with his brother William Hopper; see in particular the chapter on "Three Wise Men from the West Go to Japan".
- A book written by Lloyd Dobyns and Clare Crawford-Mason (1991), the original broadcasters of "If Japan Can, Why Can't We?", the documentary that made W. Edwards Deming a public figure in the USA in 1980. After the broadcast, the authors became aware of the activities of the CCS and made contact with Sarasohn, and their 1991 book (Dobyns & Crawford-Mason, 1991) reflects this broader understanding.
- KTKK Talk Radio, Provo, Utah. Radio talk-back interview with Martin Tanner, 18 July 1990.
- An interview with Homer Sarasohn, by Robert Chapman Wood (1989).

Acknowledgements

This paper came about because of a request for information about Homer Sarasohn that I sent to his daughter Lisa Sarasohn. In order to respond, she brought forward her plans to catalogue her father's papers for the Library of Congress, where they are now available. I am deeply grateful for all her assistance. Ken Hopper and Will Hopper were also very helpful with comment and generous with materials. I also acknowledge a profound debt to my mentors in Quality Management: Homer Sarasohn, Myron Tribus, Norbert Vogel, and Yoshikazu Tsuda, people who have helped to transform companies, industries and nations. The Associate Editor and referees provided very helpful additional information and corrections. Shu Ramada kindly provided a translation of introductory material from Homer Sarasohn's book.

References

Adams, S.B and Moranti, P.J. (2007). Global Knowledge Transfer and Telecommunications: The Bell System in Japan, 1945–1952. Enterprise and Society Advance. Access originally published online on September 20, 2007. *Enterprise Soc. 2008*, 9(1), 96–124.

Back, G.I. (1950a). Letter to Homer Sarasohn. http://honoringhomer.net/letters/george_back.pdf

Back, G.I. (1950b). Letter to Charles Protzman. http://www.puritangift.com/pdf/letter_back.pdf.

Butman, J. (1997). *Juran: A Lifetime of Influence*. New York: Wiley.

Conway, W.E. (1992). *The Quality Secret: The Right Way to Manage*. Nashua, NH: Conway Quality Inc.

Deming, W.E. (1950). *Elementary Principles of the Statistical Control of Quality*. Tokyo: Nippon Kagaku Gitutsu Remmei.

Deming, W.E. (1986). *Out of the Crisis*. Cambridge, Massachusetts: Massachusetts Institute of Technology Center for Advanced Engineering Study.

Dobyns, L. and Crawford-Mason, C. (1991). *Quality or Else: The Revolution in World Business*. Boston: Houghton Mifflin Company.

Donkin, R. (2001). *Blood, Sweat and Tears: The Evolution of Work*. New York: W. W. Norton & Company.

Fisher, N.I. and Nair, V.N. (2009). Quality management and quality practice: perspectives on their history and their future. *Appl. Stoch. Model. Bus. Indust.* 25 (1): 1–28.

Hopper, K. (1982). Creating Japan's new industrial management: the Americans as teachers. *Human Resource Management* Summer: 13–34. Republished in Japanese in *Sango Kunren* (1983), volumes 338 and 339.

Hopper, K. (1985). Quality, Japan, and the US: the first chapter. *Qual. Prog.* 18: 34–41.

Hopper, K. and Hopper, W. (2007a). *The Puritan Gift: Triumph, Collapse and Revival of an American Dream*. New York: I.B. Tauris & Co Ltd.

Hopper, K. & Hopper, W. (2007b). http://www.puritangift.com.

Howard, D. (2008). Homer Sarasohn ... The Man Who Made Japan Successful. www.firstmetre.co.uk/library documents/493.

Juran, J.M. (1951). *Quality Control Handbook*. New York: McGraw Hill Book Company.

Kitada, H. (1995). Postwar reconstruction of statistical system in Japan. In: *Forty-Year History of the Statistics Council*, 67–119. http://www.stat.go.jp/english/index/seido/snj/pdf/51ch5.pdf.

Kobayashi, K. (1986). Quality Management at NEC Corporation. *IEEE Communications Magazine* 24: 5–9. (This article also appeared in *Qual. Prog.* **19** (1986), 18–23.).

Mann, N.R. (1989). *The Keys to Excellence: The Story of the Deming Philosophy*, 3e. Los Angeles: Prestwick Books.

Moriguti, S. (1987). SQC-TQC-UQC Kyoso to kyocho no jidai he [SQC-TQC-UQC Toward the age of competition and cooperation]. *Hyojunka to Hinshitukanri* [Standardization and Quality Control] 42 (7): 32–40.

Nakaoka, T. (1981). Production management in Japan before the period of high economic growth. *Osaka City Univ. Econ. Rev.* 17 (1981): 16–24.

Noda, N. (1970). How Japan absorbed American management methods. In: *Modern Japanese Management*, 52–53. London: British Institute of Management.

Pearson, E.S. (1935). *The Application of Statistical Methods to Industrial Standardization and Quality Control*, vol. 161. London: British Standards Institution, Publication Department.

Protzman, C. W. (1950). http://www.puritangift.com/pdf/protzman_rept_48_50.pdf.

Sarasohn, H.M. (1952). *The Industrial Application of Statistical Quality Control*. (Translated into Japanese by Gonta Tsunemasa.). Tokyo: Kagaku Shinko Sha.

Sarasohn, H.M. (1997). Progress Through a Commitment to Quality. National Quality Management Conference, April 1997. Available at www.valuemetrics.com.au/resources001.html.

Sarasohn, H.M. & Protzman, C.B. (1949). *The Fundamentals of Industrial Management*. Civil Communications Section, GHQ, SCAP. 1998 electronic edition edited by N.I. Fisher, available at http://deming.ces.clemson.edu/pub/den/giants_sarasohn.html.

Taylor, F.W. (1911). *The Principles of Scientific Management*. New York: Harper & Row.

Tribus, M. (1994). W. Edwards Deming. By his works shall ye know him. A tribute to Dr. Deming. *The Community Quality Journal* Special: 4, 12, 13. Available at http://deming.eng.clemson.edu/den/nae_memorial.pdf.

Tsutsui, W.M. (1996). W. Edwards Deming and the origins of quality control in Japan. *J. Jpn. Stud.* 22 (2): 295–325.

Wood, R.C. (1989). A Lesson Learned and a Lesson Forgotten. (Japanese management learned from American Homer Sarasohn.). *Forbes Magazine* 143 (3): 70–76.

Résumé

L'histoire de la Gestion de Qualité, et le rôle que la statistique y a joué, sont inextricablement liés à la reconstruction du Japon après la seconde guerre mondiale, puis à des développements aux Etats-Unis pendant les trente années qui ont suivi. Bien que dans l'histoire des sociétés ces périodes viennent tout juste de s'écouler, il y a cependant un profond désaccord sur ce qui a été réellement fait, et sur ceux qui pourraient être reconnus pour y avoir contribué. Cet article fait appel à du materiel historique récemment rendu disponible afin de clarifier ce qui s'est réellement passé entre 1946 et1950, et en particulier la contribution d'un ingénieur remarquable, à savoir Homer Sarasohn.

Appendix: Notes on the Text

Item 1 Sarasohn recounts how he received a telegram in his office, stating "IN ACCORDANCE WITH OUR LETTER TWELVE MARCH GENERAL MACARTHERS HEADQUARTERS HAS REQUESTED YOUR SERVICES EARLIEST POSSIBLE DATE UPON RECEIPT REPLY YOUR AVAILABILITY INSTRUCTIONS FOR PROCESSING WILL FOLLOW". Sarasohn proceeded to treat this as a joke dreamt up by someone else in his office. A couple of weeks later, he received an irate phone call, asking why he hadn't responded. (The original was dated "1947 Mar 26" clearly an error.)

In the 1990 KTKK Talk Radio interview cited earlier, in Section 6, Sarasohn identified his pre-Japan experience in managing a rapid transition from prototype micro transmitters to production models as being a key reason why Douglas MacArthur brought him to Japan to establish the radio communications industry.

Item 2 This paper has been quoted extensively, as it has not been published and is available only as a web document. It has important narrative value.

Item 3 Hopper (1982, page 15) says that Professor Yoshio Kondo, of Kyoto University, holder of an individual Deming Prize for Quality, who has been active in the Japanese Quality Movement since the early post-war years, credits Magill with having first advocated that Japanese industry accept statistical quality control. Magill appears to have been in Japan for about a year. His assistance with quality control at NEC was acknowledged by Koji Kobayashi, who joined NEC in 1929 and was asked to study statistical quality control (Kobayashi 1986). Kobayashi spent his whole working life at NEC, ultimately retiring as chairman of the board and chief executive.

Item 4 The following Japanese official statistics were kindly supplied by the Associate Editor. (Note that the precise situation at the end of 1945 is unknown.)

Elementary school of 4 years became compulsory in 1900; 6 years from 1907; 8 years from 1941 before occupation.

Number of telephones:
1,618,000 in 1944
746,000 in 1945
1,192,000 in 1947
1,735,000 in 1950 (NTT)

Newspaper publication:
24,245,000 (copies per day) in 1943
15,518,000 in 1944
14,180,000 in 1945
26,848,000 in 1950

Diffusion of radio:
7,473,000 subscribers in 1944
5,728,000 in 1945
6,443,000 in 1947
9,192,000 in 1950 (NHK)

Item 5 An article entitled "Frank A. Polkinghorn, Director, 1957–1958" appeared in the *Proceedings of the Institute of Radio Engineers*, Oct. 1957 Volume 45 no. 10, *pp.* 1330–1330. It records that "From 1948 to 1950 he was on loan to the Army in Japan, where he was Director of the Research and Development Division, Civil Communications Section, Supreme Command for Allied Powers, in which position he had supervision of all communications research, development, and manufacturing in Japan. In appreciation for this work he was elected an Honorary Member of Denki Tsushin Gakkai, the Japanese counterpart of the IRE".

Item 6 Some documents state that the courses ran for eight hours each day, rather than four. However, materials from around the time of the courses (e.g. Protzman 1950) suggest that formal instruction, at least, was for four hours per day.

Item 7 In fact: "At about the same time as the CCS programs, the Management Training Programme (MTP) for middle management, and Training Within Industry (TWI) for key supervisors were initiated for business in general. MTP started as a training programme for the Japanese supervisors who were then working at Far East Air Force installations. The course dealt with a comparatively broad range of management subjects, providing instruction in management techniques required for them". Noda (1970).

In a personal communication, Ken Hopper wrote that "At the same time US airbases in Japan set up middle management training. That letter became postwar Japanese middle management training. According to my correspondence with Bunzaemon Inoue, it was like CCS training but was for Middle Managers. CCS thus had little direct influence on the eventual training given Japanese middle managers. It clearly had great indirect influence".

Item 8   In a report to Western Electric about his tour of duty in Japan, Protzman listed the courses as:

Organization Control

Supervisory Development

Engineering Organization and Control Quality Control

Budgetary and Cost Control

(Oddly, this report does not mention Sarasohn by name, saying only that "In this survey and the subsequent program of development I was assisted by a CCS radio engineer who was functional on radio and vacuum tube manufacturing".) See http://www.puritangift .com/pdf/protzman_rept_48_50.pdf.

Item 9   The earliest recognition by Deming appears to be in a hand-written note he sent to Kenneth Hopper dated 22 November 1998, in response to Hopper sending him a copy of his article "Creating Japan's new industrial management: the Americans as teachers". The note read: "Dear Mr Hopper, Your letter and article excite me. I am much indebted to you. Your article is just what I need. Sincerely yours W. Edwards Deming" (Personal communications from Kenneth Hopper to NIF.)

In a letter to Shirley Sarasohn dated 4 April 1996, William Hopper wrote "My brother Kenneth was close to Dr Deming who strongly encouraged us to pursue our investigations into the role played by the Civil Communications Section of General MacArthur's Command in Japan under the Occupation. Dr Deming even subsidized the cost of our research. When we submitted our conclusions to him, he wrote back: 'This is just what I need'".

Item 10 Deming's accomplishments go far beyond his contributions to management. The web page (http://www.amstat.org/awards/index.cfm? fuseaction=deming) associated with the Deming Lecturer Award of the American Statistical Association lists three broad areas of achievement: contributions to sampling, statistical contributions to business and industry, and contributions to management, within each of which he had great impact.

Item 11 An assessment of the work on page 109 of Butman (1997) dismisses the CCS manual as "… a cut and paste job, a synthesis of many people's ideas, cobbled together with occasional lapses of clarity, a typewritten volume filled with typos and illustrated with rudimentary diagrams", and notes that "none of them was a recognized authority on business management or quality control ... What's more, one of the seminar leaders (Sarasohn) was just 33 years old, an engineer with no experience as a senior manager in a large company". Sarasohn's annotation in a draft of Butman's book indicates that he found this a "very disagreeable and perjorative [*sic*] passage". Butman's own account is itself a "synthesis of many people's ideas" (which book isn't?) and contains numerous factual errors just in that chapter (e.g. Polkinghorn did not teach any part of the CCS Management Seminar; the Seminar ran for 4 full days per week, not 4 half days per week; Deming was 6"7" tall not 6'1", and so on).

Item 12   Japanese engineers appreciated this as well. Many years later, Tetsuro Nakaoka wrote: "SQC [Statistical Quality Control] was particularly effective when it was introduced into a production line what had already been rationalized along Taylorist lines; a typical example was the telecommunications sector. For example, Nippon Electric, after early guidance of CCS officers, had considerably rationalised its production control system with the help of the JMS [Japanese Management Association]. Then by introducing SQC methods on a large scale, Nippon Electric was beginning to turn out products of notable high quality and to be widely regarded as a model of QC (Nakaoka 1981). Sarasohn regarded this paper as "... one of the most complete and honest discussions of the subject I have seen since the end of World War II". (Letter from Homer Sarasohn to Hakim Mohammed Said, 27 April 1991.)

During his interview with Myron Tribus, Sarasohn also noted that during the latter part of the 1940s, some Japanese people felt that the Americans were refusing to teach them Statistical Process Control, as it was SPC that would help the United States retain competitive advantage. Sarasohn's counter argument was the same: first the Japanese needed to understand broader aspects of good ("quality") management. Given his ongoing debate about this issue, his own (1952) book on Quality Control, and his awareness of the work of Walter Shewhart, it is clear that Sarasohn had a good working understanding of Statistical Process Control.